Technologies in Mining

Edited by Abhay K. Soni

Published in London, United Kingdom

Technologies in Mining
http://dx.doi.org/10.5772/intechopen.111055
Edited by Abhay K. Soni

Contributors
Abhay K. Soni, Chandrani Prasad Verma, Clause Fátima de Brum Piana, Edip Taşkesen, Hamza Alahmad, Javad Sattarvand, Lizete Stumpf, Luiz Fernando Spinelli Pinto, Luís Eduardo Torma Burgueño, Manoshi Saikia, Maurizio Silveira Quadro, Maurício Silva de Oliveira, Milad Ghahramanieisalou, Pablo Miguel, Prasenjit Saikia, Samiran Bora

First published in London, United Kingdom, 2025 by IntechOpen
IntechOpen is the global imprint of INTECHOPEN LIMITED, registered in England and Wales, registration number: 11086078, 167-169 Great Portland Street, London, W1W 5PF, United Kingdom

For EU product safety concerns: IN TECH d.o.o., Prolaz Marije Krucifikse Kozulić 3, 51000 Rijeka, Croatia, info@intechopen.com or visit our website at intechopen.com.

British Library Cataloguing-in-Publication Data
A catalogue record for this book is available from the British Library

Technologies in Mining
Edited by Abhay K. Soni
p. cm.
Print ISBN 978-0-85466-815-1
Online ISBN 978-0-85466-814-4
eBook (PDF) ISBN 978-0-85466-816-8

Meet the editor

Dr. A. K. Soni, Ph.D., graduated with a degree in mining engineering from the Ravi Shankar University in Raipur, Chhattisgarh (India), in 1983. He completed his post-graduate studies at the Birla Institute of Technology and Science (BITS) in Pilani, Rajasthan, India. He obtained a Ph.D. in environmental science and engineering from the Centre of Mining Environment, IIT–Indian School of Mines (ISM), in Dhanbad in 1998. Dr. Soni is the former chief scientist of the prestigious Council of Scientific and Industrial Research - Central Institute of Mining and Fuel Research. (CSIR-CIMFR) in India. He was the head and scientist-in-charge of the CIMFR and Nagpur office and engaged in research in the mine environment and allied areas. His area of research interest is geo-hydrological problems related to mines. He has visited the USA and the UK for research work and has travelled extensively across the countries. Dr. Soni has more than 39 years of experience in research, academics, and the mining field in various areas of environment and subsurface sciences. Dr. Soni has over 10 years of administrative experience in science and R&D administration. To his credit, Dr. Soni has authored three books published by reputed international publishers, namely Springer and CRC Press/Taylor & Francis. He is an academic editor for three open-access books published by IntechOpen. Dr Soni served at Coal India Limited (at the SECL subsidiary) and the Department of Mining Engineering, Government Engineering College in Raipur (Chhattisgarh), before becoming a scientist in October 1987 at CSIR-CIMFR. His research interest has caused the development of an Environmental Degradation Index (EDI) for applications in mining in ecologically fragile areas, eco-friendly mineral extraction techniques, and improved environmental management in delicate and sensitive areas with particular reference to mining operations and the hill-type regions (Himalaya). He has a special interest in policy issues on mining and the environment. He is actively associated with technical and professional organizations and societies such as the Institution of Engineers (India), Mining Engineers Association of India (MEAI), International Mine Water Association (IMWA), and Indian Society for Rock Mechanics and Tunneling Technology (ISRMTT). He is associated with many mining and tunnelling projects in India and has been invited by Indian universities to deliver lectures and conduct examinations for post-graduate students. Dr. Soni was associated with the Bureau of Indian Standards (BIS), Nagpur University (as a member of the Board of Studies, Mining) and has experience organizing several technical events. He has delivered lectures at reputed institutes like IIT–Roorkee, NIT–Raipur, and IIT–ISM, Dhanbad. He has conducted international and national training courses, seminars, and workshops. As a research scientist and technical administrator, he has more than 120 technical publications on mining and environment-related topics in Hindi and English, excluding technical reports, which he prepared for CIMFR (erstwhile CMRI) until his superannuation on May 31, 2023. Dr. Soni has handled over 100 R&D projects as principal coordinator and principal investigator and received the best R&D paper award from a nationally accredited technical society, namely the Indian Geotechnical Society (IGS), in 2003 and the Best Citizen Award in 2006.

Contents

Preface

In order to be innovative and forward-looking, we must keep track of science and technology in any engineering field and stay abreast of them. For any mining engineer, *Technologies in Mining* is a book that entails the present-day requirement of knowing what is currently happening in the industry on technological fronts. Digital innovations and fleet management systems in mining, information technology, and wireless communication have advanced rapidly over the past two decades. Driven by smart devices, these emerging technologies are reshaping mineral mining and the industry as a whole.

The new technology concepts in mining fine-tune the industry and bring together researchers and academicians from different fields around the fundamental questions that cut across traditional disciplines. Small-scale mining (artisanal mining) and medium/large-scale mining sectors can benefit from technological advancement and development by focusing on industrial growth. Indeed, an improvement through the path of discovery and innovation can speed the growth of technology in the mines, whether on the surface or underground level. The desired industrial output is achievable by fostering collaboration and knowledge-sharing between different communities.

This book explains the application of mining science and shows how technology can benefit society and the milieu in which we live. Technology creates new devices and innovative methods to solve problems and complete tasks.

In this book, we have included chapters on different topics in significant areas of mining /excavation, i.e. mining, rock mechanics, environmental geology, eco-energy and digitization, covering a part of the technology. The introductory chapter talks about the selected key points to understand the approach of this book. The chapter "Digital Twins and the Mining Industry" is related to modern-day sensors, information, and communication technology. The chapter "Evolution of a Technosol Cultivated with Perennial Grass, over 15 Years: Potential Use as Carbon Sinks" discusses the over fifteen years of research output to know the potential uses of carbon sinks. It mainly covers climate and soil impacts in the mining areas. The research-oriented chapter "Stability of Crown Pillar during Transition from Open Pit to Underground Mining" covers rock mechanics-related aspects during the transition phase of the mine from open pit to underground. As already iterated, the book topic is vast, so in brief, varied subjects from exploration to exploitation, besides the automation part, have been covered. We attempted to include papers with a "new dimension" concerning the mining industry.

An array of clean coal technologies is scrutinized in the chapter "Eco-Energy Innovations in Revolutionizing Coal: Assessing Clean Technologies for a Greener Future", which falls within the coal technology domain, investigating their capacity to mitigate environmental repercussions and propel the advancement of a more sustainable future.

Surface coal mining is a highly impactful activity that significantly changes the mining environment and the biodiversity of the lands where it is installed. In one of the studies, the increase in carbon in the surface layer of a technosol (soil type whose development is dictated by human-created origin conditions) promoted by perennial grasses, its implications for the recovery of land degraded by coal mining in southern Brazil, and its potential as a carbon sink have been described. This research-oriented article aims to reduce greenhouse gases, especially carbon dioxide (CO_2), because new atmospheric science-related technologies have received significant attention in recent decades. The soil organic carbon in mining land is changing rapidly due to human activities, thereby significantly affecting the regional and global environment and influencing soil fertility and crop yields. Similarly, "Asphaltite Potential Assessment as Energy Source in Şırnak" explores the asphaltite that present-day mining engineers should know for exploitation and exploration uses. This chapter covers the research of Asia and Europe, as Turkey is located partly in Asia and partly in Europe.

This book adds knowledge to the reader and benefits field personnel, researchers, and other industrial functionaries. Overall, the book is a good choice for your bookshelf.

Abhay K. Soni
CSIR-Central Institute of Mining and Fuel Research,
Nagpur, Maharashtra, India

Chapter 1

Introductory Chapter: Technologies in Mining

Abhay K. Soni

1. Introduction

The topic of "technologies," as applied to the mining industry is so vast that a full book can be written. The mining industry is indeed poised for significant growth in the years ahead. Hence, cutting-edge technologies guided by discovery, innovations, and smart devices will be replacing conventional practices slowly and steadily. Various areas of technology that embrace innovations in mining are detailed in this book. The subsections of the various chapters and described modern concepts currently in use in mines, *viz.* digital mine, process-oriented opencast mine, artificial intelligence (AI), machine learning (ML), value engineering, reverse engineering for practical knowledge, wireless technologies, green mining, clean technologies for coal, quality control for coal and metal mines, automation, use of robotics and drones, block chain technology, etc., are some of the mining technologies available for the twenty-first-century mines whether surface or underground. These new engineering concepts, concerning mines of all types and sizes, are the tools and solutions that reduce the risk for hazardous mining operations. The *technology challenges,* that are ahead, have to be dealt with/by a strong policy framework and resource optimization. In this first chapter (termed an "introductory chapter"), I have written the selected and key points concisely. Based on these four/five pages, one can understand the approach with which this book is formulated.

2. What is new in mining and mining technologies?

When the word technology is discussed, it is quite often asked—what is new or latest, whether, in the mining and allied fields or any other arena of engineering? Nowhere, in a single place, all new technologies are listed or described. Depending on the subject and area, suitability and applicability are covered context-wise. In this twenty-first century for a mining engineer, a mining company and their professional's key S&T areas are mechanization of mines producing minerals and information plus communication technologies (e.g., digital innovations, data/information, and wireless technology, etc.). Broadly covering these points, given below is the checklist of topics [1] giving an overview of the latest trends and concepts for the industry as a whole. Intentionally, I have tried to keep it very catchy as this subsection of what is new can be described in many ways. Subject area wise, several of them are already inducted in the industry, and some are at the different stages of implementation. Mining technologies have a close relation with site specificity because of the characteristic the industry has. Even, some advanced concepts as well as tailor-made technological solutions are introduced in the mines with changing times. Likely, such technologies may not be listed here, but their know-how is available to industry experts.

Check list: (for the) mining and allied mineral based industry

a. Latest trends/technology concepts—*subject area: mining*

- Digital mine
- Process-oriented opencast mine
- Telemining concept
- Continuous miners and continuous mining
- Computer-aided loading of excavators
- Computer-aided mine planning, map preparations, and inventory control
- Fleet management systems in mining
- Truck-dispatch system using GPS technology and wireless
- Automation in various unit operation (from drilling to blasting, mucking to loading, including the bulk solid transportation of mineral/waste
- Technology of heavy earthmoving machines and equipment
- High degree of mechanization and automation in mines
- Robotics and drones
- Routine condition monitoring for mining
- Slope monitoring system
- Overburden/waste dump and its management
- New technology for mine safety and safety engineering
- Advanced safety systems and technology (remote operations)
- Mine system engineering
- Value engineering and reverse engineering for practical knowledge
- Wireless technologies for mine information and communication
- Quality/grade control for coal and metal mines (green and clean mining, technology options)

- Completely reskilling the workforce engaged in mines and mining operations
- Mineral marketing and E-commerce (through internet and intranet)
- Information technology for communication, IoT and IIoT, meshing and faster network development
- Edge computing and cloud computing
- Block chain technology and creation of large mining data bases

b. Latest trends/technology concepts—*subject area: mining environment*

- Artificial intelligence (AI)
- Machine learning (ML)
- Green mining
- Clean technologies (coal and metal)
- Integrated environmental management in mines
- Life cycle analysis
- Carrying capacity concepts
- Human dimension and quality of life
- Atmospheric environment vis-a-vis mine environment
- Resource modeling and environmental modeling
- Fuzzy–Logic in environmental management
- Green chemistry (for pollution control and environmental management)

c. Latest trends/technology concepts—*subject area: geology*

- Geo-Statistics (GPS/GIS technologies)
- Computer-aided evaluation of mineral and ore deposits
- Modern geophysical methods (for exploration, evaluation, and assessment of geological deposits, both surface and subsurface)
- Computerized geological mapping

d. Latest trends/technology —*subject area: ore processing and mineral conservation*

- Automation in ore processing
- Crushing, grinding, and conveying of ore (latest technologies)

- Recovery of metals from low-grade ores, poly-metallic nodules, etc.
- Plasma furnaces for smelting of finely divided ores
- Superconducting magnets for separating dry materials
- Use of automation, computer, and AI tools in mineral conservation

Thus, *Technovations* (technology + innovation), for various mines of the mining and mineral sector are meant for a better understanding of the excavation method and processes [2]. These technologies are introduced, depending on the prevalent site conditions of the mines and industry. To overcome the impediments, many challenges, general and special, are faced by the industry.

3. The industry *vis-a-vis* the book chapters

The mining, minerals, and metals industry is poised for significant growth in the years ahead. It is one of the basic industries for human civilization growth continuing for centuries. Several allied industries are attached to it providing employment and opportunities to millions of people the world over. When we discuss technology and its related aspects, words, such as science, discovery, and innovation, come into the forefront greatly affiliated with the technology. Hence, cutting-edge technologies are slowly and steadily replacing conventional practices.

Some technologies have tremendous potential to create value for the industry, its stakeholders, and society. Significant among them are as follows:

- Automation *(e.g., remote operation of the machinery and equipment)*
- Robotics (for hazardous, less safe, and underground workings applications)
- Next-generation analytics, planning, and decision support (*e.g., big data tools, etc*.)
- Integrated enterprise, platforms, and ecosystems (*e.g., GPS operation of mine fleets*)
- Digital applications for the mining industry (both software and hardware)
- New and emerging "nanotechnology" for materials, nanoparticles, and environmental systems (for mining environment-related problems and solutions).

Several of these areas are either fully grown or being developed. From the viewpoint of their implementation, the mining industry is fully geared up to embrace them. All such areas are hence emerging, new, and involve subject area expertise for candid industrial success. For their field applicability in the mines (i.e., the real ground truth), the moderation of technology is a must. Sometimes, tailor-made solutions have to be evolved, and a combination of modern and conventional practices works better, for example, a *hydrogen-powered mine haul truck* is an example of new technology replacing diesel trucks with the hydrogen model that reduces carbon dioxide emissions out of the atmosphere significantly. A hydrogen-powered mine dumper (496/290 tonners) designed to operate in everyday mining conditions is capable of carrying a 290-ton payload, which has the 2 MW hybrid hydrogen battery generating more power than its diesel predecessor. It provides a fully integrated green hydrogen system (production, fueling, and haulage) for bulk mineral/ore transportation [2]. Similarly, the information/communication technology and computer technology front guides the mining industry for practical solutions in their actual mining operations.

4. Challenges

To overcome the bottlenecks, remove impediments, and enhance the overall efficiency of both man and machinery, the technological challenges could be either long term or short term. Globally, the mining industry is capital-intensive and labor-oriented. Hence, among the various challenges, one significant challenge that worries the government and companies is *employment*. Technology has both dimensions, that is, additional employment generation (indirectly for other sectors where deployed) and curbing employment because of automation. Therefore, "capital and cost," both are challenges for the industry. They need focused debate, whenever a technology enters the industry. The skill/skill set and generation gap matter a lot in dealing with the capital and cost challenges faced by the mining industry. In this regard, *Industry 5.0 Vs. Mining 4.0* and emerging global trends in the mining sector is a quick and short answer. It deals with the operational challenges, environmental challenges, and climate changes as well [2].

5. Policy strategy

Policies declared to date for the mining and mineral industry globally recognize the need for sustainable, environmentally friendly, and cost-effective management at every stage of the mining project from exploration to exploitation [3]. An integrated strategy that links the technological options with the national mineral resource base has to be understood prima facie with the changing paradigm (**Table 1**).

Past/present	Future
Mining techniques are a mix of traditional and modern methods	Eco-friendly, remote-controlled, and AI-based clean technology options; E-commerce
Medium to low standard of living in mining areas besides large influx of money (poor infrastructure/civic amenities/transportation/information and communication/energy and power utilities	Raising of quality of life/standard of living in general in mining areas with improvement in modern amenities and information technology
Age-old, orthodox, and typed culture in the mining belts	Cultural changes as per modern developments
Comparatively less strong quality human resource base in both technical and nontechnical cadres	Planned policies to attract strong and skilled human resources
Lack of innovation in far-flung mining areas	Quality-based technical innovations
Dull image of industry (no white-collar jobs, pollution, and remote location-oriented mine and mining areas)	Better image projection
Easy to operate mines in interior areas	Social license to operate; rural-urban interface for sustainable development

Table 1.
Changing paradigm for the mining industry [4].

Science and technology when get married, which are impossible to separate, the role of policy strategy comes into the picture. A sound industrial policy, with past, present, and future strategy encompassed, leads toward industrial advancement. One of the difficulties, mining companies or mine management focuses, with regard to policy formulation and implementation is the lack of proper machinery and equipment, expert knowledge, and financial resources. Hence, the gaps which exist can be solved effectively through policy intervention [3].

6. Conclusion

In conclusion, this introductory chapter provides readers, with a basic overview of the *technologies in mining* that revolve around the technology-driven exploitation techniques for minerals. The thoughtful contributions made by a number of the author(s) have enhanced the scientific rigor and great value in totality. The research input and content of the book, depth in methodology, interpretation, and discussions have certainly addressed the burning issues of the industry. Hence, I will recommend that readers glimpse at this book.

Speedy mineral production and an accelerated demand in the twenty-first century are witnessed because of population growth. Hence, it is rather difficult to run industry without technology. Utilization of knowledge of mining engineering [5] and human dimensions of environment [6] together with technology can benefit to the industry and turn the filthy mining operation into a green and clean industrial operation. Therefore, it is correct to conclude that our scientific learning be technology-driven to achieve the desired goal/result. I will suggest and recommend that readers glimpse at this book.

It is true that, augmenting sustainable development in the mining industry mines and their various unit operations, including mine environment problems, need to be handled scientifically and technologically. Technologies of the modern day are well equipped to felicitate it. The future direction of the technologies in the mining industry including its adoption, being dealt with by the stakeholders, should be progressive and implementable, along with their user-friendly (easy to use) and cost-effective characteristics. It should match with the regulatory framework of the country too where these are implemented into practice. Technologies that cause the simplification of working procedures and expedited for more earnings and profit for the company are considered as a boon.

Author details

Abhay K. Soni
CSIR-Central Institute of Mining and Fuel Research, Civil Lines, Nagpur, Maharashtra, India

*Address all correspondence to: abhayksoni@gmail.com

References

[1] Soni AK. What is new in mining, mine environment and mineral conservation? In: Souvenir of 10th Mines Environment and Mineral Conservation Week—2001-2002, Dehradun Region. H.P., Delhi, U.P., Haryana, Uttaranchal. 2002. pp. 7-9

[2] Soni AK, Muthreja IL, Yerpude RR. Principles and practice in mining engineering. In: Emerging Areas of Technovations in Mining. India: CRC Press, Taylor & Francis; 2024. pp. 225-260 (A Balkema Book) (Paper-back Edition; ISBN-97810032228198, Pages-350)

[3] Soni AK, Wolkersdorfer C. Mine water: Policy perspective for improving water management in the mining environment with respect to developing economies. International Journal of Mining, Reclamation and Environment. 2015;**30**(2):115-127. DOI: 10.1080/17480930.2015.1011372

[4] Soni AK. Development of mineral sector and mineral based industries in Uttaranchal. In: All India Seminar on Infrastructural Development in Uttaranchal-Problem and Prospects, October 11-13, Institution of Engineers, Roorkee Local Centre, Roorkee. 2001. pp. I–31-I–34

[5] Soni AK, Dwivedi RD. Utilization of knowledge of mining engineering in underground civil construction work. Mining Engineers Journal. 2002;**4**(1):22-23

[6] Soni AK. Human dimensions of environment. In: Souvenir of XXI Mines Safety Week—2001, Saurashtra Region, November 26–December 02, 2001. Veraval (Gujarat), India: Gujarat Ambuja Cements Limited; 2001

Chapter 2

Digital Twins and the Mining Industry

Milad Ghahramanieisalou and Javad Sattarvand

Abstract

Safe and efficient assimilation of new technologies into current operations in the mining industry requires adapting to new challenges. Traditional mining techniques and operations will inevitably be adjusted to incorporate new methods and machinery. Various industries, from manufacturing and engineering to social sciences, have embraced the Digital Twins (DT) methodology to study complex systems. The benefits of DT, encompassing features like a data hub, simulation and analysis tools, and visualization platforms, are substantial because they replicate their physical counterparts even before their existence. Once the physical twin is constructed, the DT serves as a digital mirror, aiding in ongoing monitoring, improvement, and control. Digital Twins utilize data-driven and physics-based models and advanced analytics to optimize cost, environmental emissions, and resource usage in developing extraction, production, processing, refining, manufacturing, or recycling technologies. They also enable precise control, predictive maintenance, and identification of potential bottlenecks or inefficiencies through simulation, monitoring, and analysis of every step in the supply chain. Utilizing digital twins expedites the development of novel technologies, ensuring their sustainability and competitiveness. Moreover, digital twins could play a role in diversifying commercially viable and environmentally sustainable sources of critical materials, including their recovery from waste streams.

Keywords: mining industry, digital twins, simulation, cost savings, sustainability, critical materials

1. Introduction

The Digital Twins (DT) methodology has garnered significant attention in recent years, finding applications across various industries. DT offers a profound understanding of complex systems, making it especially relevant in the context of emerging technologies in the mining industry. The assimilation of these technologies while ensuring safety and the potential changes they bring to mining techniques necessitates adjustments in current operations. Predictive simulations play a pivotal role in comprehending unforeseen scenarios and facilitating cost-effective modifications, transitioning them from the operational phase to the design phase.

The value of digital twins in business lies in their capacity to provide a comprehensive understanding of processes, facilitate optimal decision-making, and monitor practical actions. Digital Twins offer insights into both real-time and simulated behavior of physical systems, leading to quicker decision-making. As a unified hub for various data formats and sources, DT integrates valuable information to enhance awareness based on internal and external events necessitating appropriate actions. These factors include people (competitors, customers, legislators, or suppliers & supply chain), equipment breakdowns, process failures, and weather events.

DT is a dynamic concept that relies on data availability, resources, and managerial decisions [1, 2]. Notably, Negri et al. [3] contend that DT represents an evolved version of the Virtual Factory (VF). The key distinction lies in DT's ability to incorporate realistic synchronization, leading to informed decisions regarding current and future production. The data model for DT must encompass system operations, historical data, behavioral patterns, and current states.

Many researchers view DT as a tool for maintenance, diagnostics, and prognostics (health analysis). The ultimate goal is to create a model that faithfully represents the complex system's behavior, accounting for external factors, human interactions, and design constraints. Improved situational awareness by DT is based on real-time or near real-time data flow from diverse sources, such as web services, information from IoT devices, or smart device machine-borne data. The fundamental concept of DT involves understanding the current situation, projecting the future, and acting accordingly. Augmented by predictive models and analytics, DT becomes prescriptive, allowing for comprehensive decision support and autonomous actions [4].

Embracing DT for "real-time geometry assurance" in manufacturing product design, Söderberg et al. [5] propose its utility in predicting undesirable outcomes and establishing a tangible link to the physical system. This connection, maintained throughout the lifecycle through sensors and actuators, enables managers to monitor the machine's current state, adjust settings, and reduce downtimes.

DT can potentially optimize mining operations across four key stages: Design, Pre-operation, Operation, and Rehabilitation. Historical data is harnessed for simulations during the design phase to identify system deficiencies and environmental interactions. In the pre-operation stage, DT aids in planning an effective monitoring scheme, determining data acquisition points, and specifying real-time or near real-time data requirements [6, 7].

The actual benefits of DT become increasingly evident during the operation stage. Continuously updated with precise data, DT projects the physical system's status and serves as a simulation environment for exploring potential future states of the mine [8, 9].

One of DT's significant strengths lies in integrating readings from all internal and external sensors and modeling the system and its surrounding environment. This capability proves critical in the final stage, rehabilitation, where environmental concerns in the mining industry are rising. DT provides a comprehensive history of mining operations and a platform for evaluating various rehabilitation scenarios [10, 11].

The emergence of DT hinges on the real-time synchronization of field data facilitated by the Internet of Things (IoT). IoT encompasses electronics, software, sensors, and network connectivity that gather and transform data. These "things," or devices, are intelligent, embedded, networked systems within production facilities. Developing information models (structures ensuring digital data continuity) is crucial for harnessing IoT to its fullest potential [3, 12].

IoT devices facilitate the transfer of information from the real world to the virtual realm (DT). If employed for operational control, DT can return the results of any analysis and simulations on the received data to human decision-makers or employ AI to issue control orders [13–16].

The business impact of DT spans increased revenue, reduced costs, improved customer and employee experience, compliance with rules, and risk reduction. DT acts as a catalyst for digitalization projects, initially aiming to improve efficiency but later evolving into monetizing new digital assets. From a societal standpoint, DT contributes to cost and time savings, job creation, new skills development, reduced traffic congestion, and a safer working environment. Environmentally, it leads to lower CO2 emissions and improved monitoring of wastes and tailings [4].

DT also changes business processes and technological advancements, with initiatives like Asset Administrative Shell (AAS) and Digital Twins Definition Language (DTDL) focusing on DT's technical definition and instantiation [4]. These are essential initiatives since the existing literature offers various (sometimes contradictory) definitions of digital twins. DT's definition and purpose depend significantly on the specific applications and vary from case to case. However, after thoroughly examining the existing academic and industrial documents, few DT definitions were found to best apply to the mining industry (**Table 1**).

However one decides to define it, the most crucial key to the DT concept is a synergistic multi-way coupling between the physical asset, data collection, computational models, and the decision-making process [23–26].

According to M. Grieves & Vickers [17], DT could be used in all of the lifecycle phases of the physical asset. To realize the evolving nature of the DT model and its full utilization, DT must be created in the design phase, used in the production of its physical twin, provided support during its operation, and stored all information in the disposal stage [27].

For this purpose, the terms DTP (DT Prototype) and DTI (DT Instance) are defined to detail DT's application in various sections of lifecycle management. Like any other prototype, DTP is used in the design phase (no physical asset exists) as a testing platform to ensure that the proposed design meets all the requirements on tolerance limits and safety features while analyzing it in different scenarios [17].

Definition	Ref.
Digital twins (Product Lifecycle Management (PLM), Mirrored Spaces Models, and Information Mirroring Models) are defined as a set of virtual information constructs that fully describe a potential or actual physical manufactured product from the micro (atomic level) to the macro (geometrical level).	[17]
An integrated multi-physics, multi-scale, probabilistic simulation of a vehicle or system that uses the best available physical models, sensor updates, fleet history, etc., to mirror the life of its flying twin. It is ultra-realistic and may consider one or more critical and interdependent vehicle systems.	[3]
A coupled model of the actual machine that operates in the cloud platform and simulates the health condition with integrated knowledge from both data-driven analytical algorithms and other available physical knowledge.	[18]
Virtual substitutes of real-world objects consisting of virtual representations and communication capabilities make up smart objects acting as intelligent nodes inside the Internet of Things and services.	[19]
The simulation of the physical object itself to predict the future states of the system.	[20]
An integrated multi-physics, multi-scale, probabilistic simulation of an as-built system, enabled by Digital Thread, that uses the best available models, sensor information, and input data to mirror and predict activities/performance over the life of its corresponding physical twin.	[21]
Digital twins are virtual representations of physical assets enabled through data and simulators for real-time prediction, optimization, monitoring, controlling, and improved decision-making.	[22]

Table 1.
DT definitions in the literature suit the mining industry.

During the operation stage, any physical system shows "predicted and unpredicted" behaviors; some of them are "desirable," and others are "undesirable." DTP could help point out the UUs ("unpredicted undesirables") in the design process, enabling modifications to the system. Also, the general types of these UU behaviors could be identified, which helps incorporate protective measures into the system. Unlike physical prototypes, DTP failures are not costly and are not a "one and done" situation [17, 28].

On the other hand, a DTI (Digital Twins Instance) would model its physical twin's as-built condition for its lifecycle (from production to operation and disposal). The aim is to monitor the state of the product for diagnostic purposes and provide predictive insights into a range of possible future states. Also, at this stage, DT will continuously evaluate the results of DTP simulations and DTI predictions to evolve itself [17, 29].

As M. Grieves & Vickers [17] pointed out, the final disposal phase has not received as much attention as it must. Systems may retire, but their information could be retained and used later on a similar project at a meager cost. Also, as for sustainability efforts, DT offers a view of the entire history of the system so proper disposal can be planned without harming the environment.

2. Digital twins' potential within the mining industry

Historically, the mining industry has always collected large amounts of data; however, there is a significant lack of effort and resources to analyze them. One of the critical obstacles to data analysis is the lack of communication and integration between the wide range of mining operations [30].

Exploring the correlations of seemingly independent processes and variables through data-driven models will enable managers and engineers to create a holistic approach to ensure optimized production. Admittedly, the quality of recorded data (past or present) has a significant role in the efficacy of this approach. However, a more pressing issue is the availability of representative data. Furthermore, data quality is a very fluid concept depending on specific points of view (noise, corruption, bias, etc.) [31, 32].

This section analyzes DT's potential in mining engineering based on DT's applications in various engineering and manufacturing disciplines. For instance, fleet management, an already complicated issue, has become even more elaborate with the introduction of autonomous or semi-autonomous machines to the mine site. This integration introduces problems such as unique health and safety issues and regulations, resource efficiency, human-machine interaction, worker training, maintenance optimization, and many more [15].

By considering the enormous expenses and risks related to a pilot project, DT could replace that by acting as a prototype in which realistic tests and simulations are carried out with close to zero cost. DT includes a data hub, simulation and analysis tools, and visualization platforms to enable appropriate designs and monitoring plans focusing on unknown areas. A few examples of DT's application in mining include fleet management, Mine-to-Mill optimization (especially D&B), geotechnical digital twins (gDT), etc. [33]

When developing regulations, testing the efficiency of the mine layout, design, and magnitude of the restricted area and its access points, human-machine interactions, risk identification, and management processes, examining the efficacy of control systems, and monitoring, DT could help speed up (or completely replace) the process in the physical world.

By examining the related literature, the potential application of DT in the mining industry is divided into 1) Geotechnical, 2) Drilling and Blasting, and 3) Fleet Management.

2.1 Geotechnical DT

The challenge of rockfall, slope instability, tailings dam failure, etc., and their associated risks have haunted the mining industry for quite some time. It has been a significant cause of fatalities, serious injuries, and financial losses. Following the established empirical techniques is the most efficient (probably the only) approach to studying this problem. But they have their disadvantages; chief among them is the thorough and resource-exhausting geological studies [34].

Recent technological developments and their introduction into mining engineering applications open the door to updating the studies. Such renovations would mainly focus on automating geological mapping of the rock surface using unmanned aerial systems and big data analysis. Also, efforts will be taken to model the potential trajectory of the fallen rocks in a semantic 3D model based on machine learning algorithms [35].

UAV photogrammetry and LiDAR were used to gather point cloud data to build a 3D model of the rock slope and extract its geological features. Furthermore, a trajectory estimation model of rockfalls, also developed at UNR's Mining Automation lab, uses the mass and origin of rockfalls to calculate the impact characteristics and simulate a rockfall's energy changes during its fall [36, 37].

Both these developments rely on a combination of physics-based and data-driven models, necessitating continuous data flow. Moreover, the mine site is a dynamic environment, and changes happen daily; therefore, the information about the rock surface must be updated at timely and financially efficient intervals.

The empirical methods used for generations are our best bet at calculating the rockfall probability; however, the possibility of improving these techniques is also explored by using machine learning to extract correlations between rockfalls and the factors that affect them. One major challenge is the availability of historical data on rockfall cases.

Geotechnical digital twins of the mine site, capable of combining a high-fidelity 3D model of the site with semantic data received from monitoring, lab tests (including coefficients of restitution), and simulations, will present all this information in a single unified interface for interactive decision-making and developing a rockfall risk map.

2.1.1 The framework of "gDT"

Fan et al. [38] have explored the applicability of DT in disaster management. In their case, DT is used as a unifying platform for all the crisis information, to implement AI in the situation analysis, decision-making (including resource allocation), and cooperation of different parties; and to better understand the interactive effects of decisions and actions in disaster management. **Figure 1** shows their ideology for "disaster city DT"; the first step is data acquisition. Disasters affect the data-gathering devices, too, so it would be possible to address the data depravity of disaster situations through AI and remote sensing technologies (as well as crowdsourcing and social sensing). Data will come from different sources; therefore, the next step is to integrate them (using AI algorithms like knowledge graphs and network embedding) to draw as much information (through ML) as possible. DT can help with the training and cooperation of responders in disasters; this

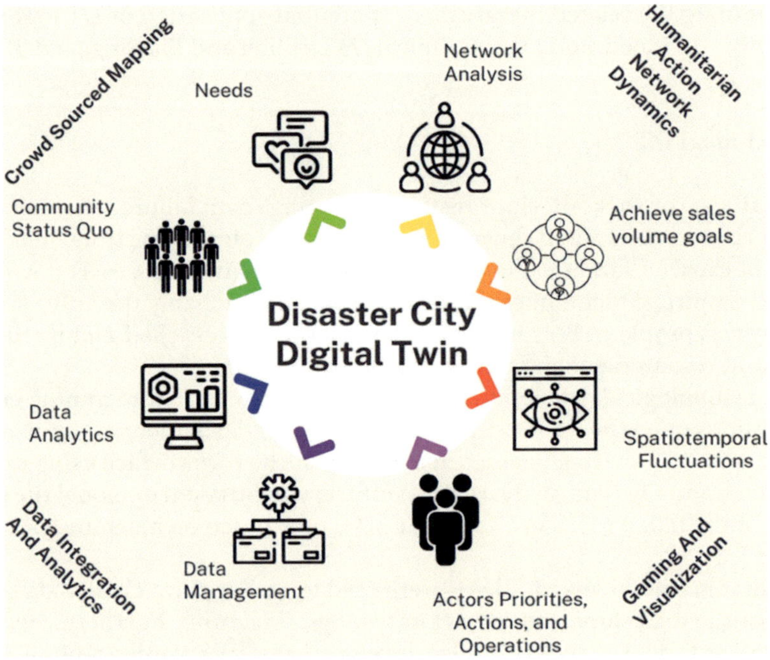

Figure 1.
Disaster City digital twins components [38].

capability is realized through "serious gaming environments" that also provide visualization. Analyzing the interaction of responders and the information they need would help with resource and task allocation. The iterative process allows the system to learn, grow, and provide predictive simulations.

Typical sources of disaster information include remote sensing technologies such as satellites and UAVs, which provide images and point clouds through photogrammetry or LiDAR. UAVs are particularly useful in disaster situations as they are effortless and require little infrastructure support [38].

Data integration is probably the most challenging part of the DT system. Fan et al. [38] propose using knowledge graphs (as heterogeneous information networks) to address the issue by defining central identities and network embedding. Data integration effort would lead to an enriched deep learning algorithm for DT to characterize the contents of images and sensory information while providing predictive simulations.

Networked analysis (using the Bayesian network) within DT helps visualize each actor's knowledge, tasks, and relationship with others (priority) to provide them with required additional data, resources, and cooperation opportunities [38].

Investigating a smart city's resiliency in case of natural and human-induced disasters, Losier et al. [39] argue that DT coupled with immersive technologies would enable disaster management first to examine the city's preparedness in case of an emergency using simulations and second to analyze the effectiveness of preventive or mitigating strategies, plans, and technologies.

The financial losses and fatalities of natural disasters have increased in the past two decades. Critical infrastructure (transportation, power, and water supply systems, hospitals, etc.) not designed for extreme conditions are most vulnerable in disaster situations. An example would be the power plants built near shorelines for water access [39].

Losier et al. [39] report that predictive simulations, mitigating technologies, and strategies are the most effective tools in managing any disaster event. For instance, practical forecasting efforts would result in a 35% reduction in annual flood damages, or highly resilient critical infrastructures (e.g., airports and power plants) would seriously limit or even prevent the consequences of an extreme event.

Focusing on the loose bits and pieces in a construction site that could become airborne during extreme winds, Kamari & Ham [40] have also examined DT's application in disaster preparedness. The as-is condition of the construction site based on the point cloud data generated from UAV photographs helps identify and locate potential health and safety hazards. Threat identification is crucial as it is the prerequisite for mitigating efforts and training workers in simulated environments.

Point cloud (PC) segmentation clusters objects together, and an AI algorithm determines whether that object has the potential to become airborne or not. The semantic segmentation of PC data is either data-driven (which relies on trained datasets using supervised learning) or model-driven (based on hand-engineered cost functions that work like shape fitting or region-growing algorithms). Model-driven methods are computationally simple but do not produce robust results concerning geometrical variances and noisy PC data [40].

Kamari & Ham [40] use the SfM (Structure-form-Motion) framework in combination with Multi-View Stereo to generate the site's 3D point cloud. The primary wooden material and its boundaries are identified in 2D images in the next step. Based on the information from 2D images, semantic segmentation (Convolutional Neural Networks) is performed on the location of identified objects using the camera's intrinsic (focal length, distortion, etc.) and extrinsic (orientation and location) parameters. One of the shortcomings of this system is its inability to identify occluded objects (due to shading), which can be solved by developing a depth map from the different camera locations.

The RANSAC (Random Sample Consensus) algorithm is used for ground registration (plane fitting) based on segmented point cloud data. The point cloud data is then transferred to a cartesian system so that the z-axis would be parallel to the projection direction. A bounding box encloses all the pixels at the XY plane, and the average height of all the points representing that pixel is calculated. To compute the volume of objects, the square size of pixels in the bounding box is multiplied by the summation of average height values [40].

To calculate the risks associated with every identified object (or pile of objects), Kamari & Ham [40] use kinetic energy and critical wind speed (min. wind speed to lift objects into the air) based on object shape, size, and characteristics. Therefore, the wind speed is critical to decide whether an object would become airborne; at any given wind speed, objects are assigned different risk levels that determine priorities for preemptive measures to control loose objects.

Many mine sites are already using radars and LiDARs in addition to more traditional in-situ instrumentation to monitor the structural stability of the mine to predict instability-prone areas.

DT provides a realistic 3D model of the mine environment, which is not only a powerful visualization tool with accurate measurements (like slope height, angle, and bench width) but also contains semantic data about the geological profile (strata and features like faults and joints) which can be obtained from probe holes and automated surface mapping, by integrating all the data from monitoring sensors (both in-situ and remote) and having an up to moment knowledge of the structural health of the mine. One of the most valuable features of DT (considering the empirical nature of stability analysis in mining) is its ability to absorb historical data from the site under study and similar projects, enabling predictive simulations and dynamic decision-making.

Considering the similarities of modeling challenges and the lack of publications directly related to the mining industry, the literature on 3D modeling efforts in civil engineering projects is examined [41, 42].

An instance would be bridge modeling, which will later be used for operation and maintenance (O&M) purposes. There are multiple challenges when creating a 3D high-fidelity model for existing bridges. If the model is based on point cloud data, even though existing techniques could cluster points for each distinct object, building a 3D shape that would fit these clusters is not automated. It takes a lot of time because of the irregular shapes of objects (95% of the whole modeling process) [43].

Furthermore, the existing methods cannot quantify the spatial accuracy of resulting standardized descriptive models (i.e., Industry Foundation Classes). R. Lu & Brilakis [43] propose a slicing-based technique that generates the "geometric DT" of an existing reinforced bridge. Assuming proper object detection, the method would take labeled point clusters as input, extract geometric features, detect shapes, and fit an IFC object to each of them (features and shapes). Also, the Level of Detail (LoD) determines the precision of the model based on its application and requirements.

The generated 3D model is transformed into a point cloud to quantify the accuracy. The result is compared to the original point cloud data using "cloud-to-cloud" distance-based measurements. The results of implementing this method on the point cloud data of 10 bridges show an average modeling distance of 7.05 cm (while the manual methods achieve 7.69 cm). Still, it is much faster than current practices (modeling time of 37.8 s) [43].

Another example of DT's application in bridge maintenance is the Bridge Maintenance System (BMS) introduced by Shim et al. [44], which uses a combination of UAV photogrammetry (for lateral and top surfaces) and Laser Scanning Cloud Data (for the bottom surface). BMS introduces the possibility of combining different techniques for gathering heterogeneous information and data from each physical asset or system component, modeling them, and combining them in the DT environment. They have also used image processing for crack detection and have demonstrated how the 3D model could be updated after each analysis to represent the up-to-minute as-is condition of the bridge.

H. Zhang et al. [45] have proposed a DT-based monitoring system for geological hazards. The evolution of the prediction methods for such hazards has led to a dynamic system capable of "real-time" analysis and prediction (using a Back Propagation Neural Network to calculate the probability of a hazard). Combining data from GIS, GPS, and various remote sensing technologies (indicating the status of the topography) with historical data (landslides, subsidence, collapses, and meteorology) provides a data visualization tool (based on GIS) and a predictive simulator of the behavior of the slope. Another advantage is that monitoring shortcomings would be identified and addressed using this system. This system would not only warn all the vehicles and people near unstable slopes (through the internet and the internet of vehicles) but also provide the opportunity to conduct mitigation and control strategies (**Figure 2**).

As mentioned above, data fed to the DT includes environmental data, historical disasters and related information, and the vehicles in the area (through a monitored checkpoint) [45].

The visualization part of the DT is also used to guide and warn all the devices and vehicles in the area and the first responders. Based on this information, it would be possible to design appropriate monitoring plans focusing on unstable areas [45].

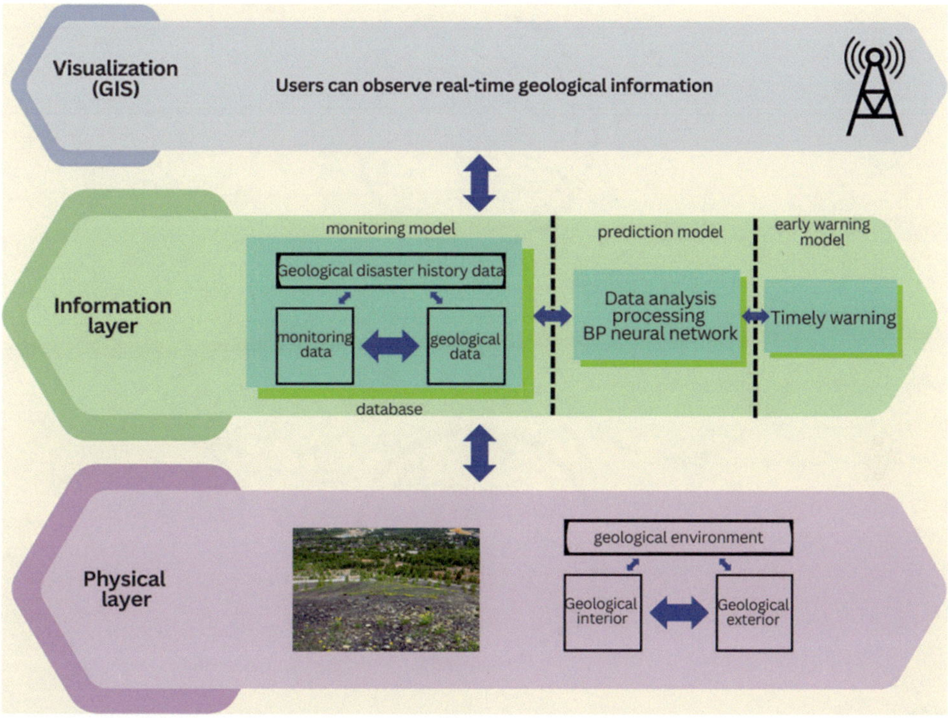

Figure 2.
The DT method is used for geological disaster warnings. Inspired by ref. [45].

2.1.2 Risk analysis capabilities of "gDT"

Emergencies and accidents are an inseparable part of the mining site, and DT could help organize the first and secondary response efforts. "Smart City Digital Twins (SCDT)" is an idea developed by Ford & Wolf [46] for disaster management. They divide the disaster experience into four steps: mitigation & preparation (happening shortly before and during a disaster), response (within days and weeks), and recovery (months or years after the disaster). The experience is unique for every community and depends on the severity and the type of disaster, which demands specific plans for every community that would consider its characteristics.

As Ford & Wolf [46] point out, a disaster would destroy individual components and their interaction with each other. DT must be able to gather and combine information from the Smart City sensors first to visualize the current condition to help with emergency response and second to predict disaster evolution and the effectiveness of the response strategy.

Figure 3 shows DT's iterative method to help with response optimization. Data is collected, response strategies are planned and tested, and decisions are made (based on the simulations), which leads to actions. The actions would affect the community; therefore, the model is updated, new plans are made, and the iteration continues until the desired state is achieved [46].

The most crucial difference between traditional disaster management and the proposed SCDT system is the close loop of data, decisions, actions, and condition updates. Also, the proposed method can model the disrupted system's components and interactions. To realize the true potential of SCDT, the system must be able to

Figure 3.
The iterative response model is used by SCDT [46].

conduct near real-time data acquisition, analysis, and simulations to provide timely feedback for decision-making [46].

In addition to all the benefits of SCDT in disaster management, Ford & Wolf [46] raise another compelling reason for developing SCDT for disaster management. They argue that as disaster management requires specifically defined information and is not concerned with the fidelity of the model (thus reducing the modeling time), it would be a great starting point and testbed for developing SCDT to its full potential: city planning and management.

The SCDT system has three types of components: parts that are the subject of monitoring and prediction efforts (shaded areas), sensors (data acquisition), and

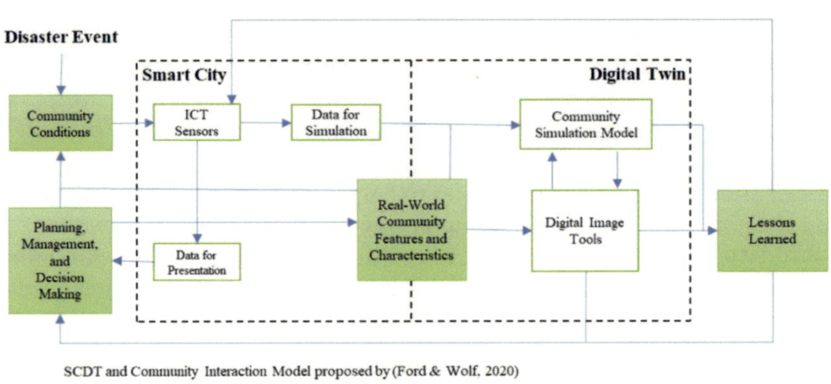

Figure 4.
Schematic of the SCDT system and its interaction with the community [46].

visualization and prediction models (**Figure 4**). After each iteration, the effect of the decision on the community's "condition" (the availability of resources after a disruption) and "features" (overall infrastructure systems capacity) is assessed [46].

The monitoring data are essential in determining the potential hazards and the efficiency of mitigation measures installed at the site. Having all the data in a centralized space will enable further study of the risk factors and their effect individually and collectively.

2.2 Drilling and blasting DT

Mine-to-mill optimization depends highly on the efficiency of drilling and blasting (D&B) practices as it affects the downstream mineral beneficiation process. The research team at the University of Nevada's Mining Automation lab investigates the potential of taking advantage of recent technological developments and their introduction to mining engineering, such as exploring the application of the Digital Twins (DT) methodology in D&B analysis. The innate complexity of D&B practices requires an equally universal analysis approach, considering the number of variables affecting it. The existing analysis techniques are based on simplified physics-based equations or data-driven models. However, DT can merge them to offer the universality and accuracy of data-driven models and the interpretability of physics-based models. Data is collected in several stages of D&B, including but not limited to measure-while-drilling data imported from smart drills, fragmentation data, crushers and mills performance, blast site characteristics, charging and timing information, blast design and geology, hole-logging, and laser-scanned profiles, drone images/videos, and operational KPIs. In a DT framework, all the data collected during D&B operations and downstream are linked under a common platform. Through data-driven modeling that depends on AI and machine learning, operators can begin to see correlations between different parameters that were unnoticeable using other modeling methods. Integrating these models into existing empirical or physical simulators would also help address the black-box approach of data-driven models. Once the DT model is formed, D&B designs will be analyzed in real time, and predictions on their performance (fragmentation, back break, fly rock, and air overpressure) will be made based on physics laws and historical data. In the day-to-day decision-making process, a D&B engineer could rely on DT for performance evaluation of the D&B program at their mine and their design capabilities over time. Various stakeholders can benefit from DT as it becomes a learning tool that can help guide their operations to be more consistent and predictable to produce repeatable desired D&B results.

2.2.1 The framework of a drilling and blasting DT

A drilling and blasting DT must be able to capture blast information and analyze it using the already existing key performance indicators (KPIs) to search for the factors affecting the outcomes by delving into the unobserved physics of the process.

The first step in realizing the DT is to understand the potential data sources; rock characteristics and bench design plans, drill rigs and supporting equipment, explosive material, and personnel are among the essential components controlling a D&B session.

The drilling and blasting DT will be a hub for a high-fidelity 3D model, including the rock mass's geological features and drill holes, to offer a powerful simulation tool providing a holistic overview of the rockfall study and analyses.

Simulations will be based on a combination of data-driven and physics-based models to reduce uncertainties while improving the interpretability of the results. Importing historical data into the system and complementing it with state-of-the-art technologies makes it possible to make an objective approach capable of producing reproducible results.

The type of rock and minerals, clay content, water content, structural features like faults and joints, etc., must be included in the high-fidelity 3D model of the mine environment used in a DT. The current methods for acquiring this information must be analyzed further to understand their compatibility with each other and the DT methodology. Geological and geotechnical information and resource estimation data are produced and stored digitally using block models. These block models can be tied to the digital twins to update as more information becomes available dynamically. Based on this information, any design scenario is stored as CAD files that can be used by planning software and easily transferred to smart drill rigs in the field.

Drill rigs will be essential to a drill and blast digital twins from a data acquisition perspective and controlling the machines themselves. Smart drill rigs need input data (drill hole design patterns) from DT to work correctly, and at the same time, sensors installed on the machine will collect and send valuable information back to the DT. There is also a significant amount of geotechnical data that the drill rig can facilitate. The location of the drill head using GPS transmitters (with connected ground stations for accuracy), rotary speed and torque, compressed air pressure and flow, vertical and horizontal vibrations in drill string and mast, drill level, and mast angle with the vertical are just a few examples of the machine data [47].

Measure-while-drilling data, such as current drilled depth, feed force exerted, and rotary torque, which is common these days, will also help track the progress of data in a real-time manner, which is ideal for DT's definition [48]. DT can use these measurements to determine secondary parameters such as rate of penetration and blast ability index. UNR's mine automation lab has also produced a system that can use drone imagery to analyze the location of drill holes and compare them to the as-designed CAD file so the engineers can devise mitigating measures in loading and sequencing. This system's advantage compared to the GPS is that it would work fine even on the deepest benches and would not need any ground stations.

Loading drill holes with explosives probably need more attention regarding digitization. It can be analyzed in the explosives and the loading process. The type of explosives used depends on the bench and rock conditions, the type of blast and its purpose, and the number of explosives used. Explosive consumption for every bench is well documented through blast reports and magazine inventory records [49]. This information must be digitized and stored within the DT. Safety measures are always the priority in any D&B stage, intensifying in the loading step. Safety records are also archived to ensure personnel safety and demonstrate adherence to regulations [48]. These records can be in analog or digital form, but they are well maintained, and their complete digitization could bring tremendous value to DT modeling and simulations.

During the loading process, one of the significant steps is to check for "hot holes" to avoid charging drill holes with high temperatures (+131°F). Remote moisture measurement and surface temperature analysis using UAVs could help [49]. Verifying each drill hole's depth and the water level is a crucial step in designing the profile (stemming column's length, the type and density of explosives, and air decking) that will be used to load the drill holes. All this information is recorded digitally on the powder truck and by the blaster [48].

Several instruments, such as high-speed cameras and seismographs, measure specific parameters (burden measuring, blast vibrations, fly rock, and air overpressure) during blasting and can determine how well the blast has been executed [47]. Other secondary parameters can be calculated within DT to analyze further a blast concerning the explosives used, including the powder factor.

After the blast, trusted KPIs within the industry are used to analyze the whole process and the efficacy of blast design.

Rock fragmentation is probably the most telling performance indicator, but the current methods to measure and determine this number are not always as efficient as hoped.

Several techniques can extract the fragmentation data in a much more efficient way. Among these are photographic methods to extract particle sizes in the muck pile, crusher monitoring, secondary blasting/breakage costs, and shovel monitoring. The crusher energy consumption, the size of the crushed product, the size and strength of the feed material, and the crusher throughput are collected from cameras and sensors installed in the machinery. As for the shovel, diggability is primarily based on the crowd armature voltage and current, the hoist armature voltage and current, the dipper trip relay, and the crowd/propel relay [47, 48]. Together, this information can help assess the fragmentation outcome of a blast inside the DT.

In the proposed framework, data from the physical twin and its components to the DT gets stored, processed, and analyzed. Based on the results of those analysis efforts, control orders are issued to control the process in the real world. To comply with industry needs, performance analysis will be mainly based on KPIs widely used by field engineers and managers worldwide.

As discussed in the previous section, by dividing KPIs into their corresponding stages of a D&B session, DT's analysis of the process would be straightforward, making it easier to find the source of potential shortcomings. For instance, drilling performance could be analyzed using collar deviation, and down-the-hole variation could help engineers track the as-drilled status of drill holes and make the necessary adjustments before the blast and for future designs. To better understand the reasons behind these errors, parameters such as rate of penetration, drill bit life, gallons per feet drilled, gallons per engine hours, engine hours per compressor hours, and feet drilled per day will help distinguish between ground and machine's effect on deviations from as-designed plan.

2.3 Fleet management

2.3.1 Human-machine interaction

One of the main areas of DT applicability is examining human-machine (human-robot) interactions. Malik & Bilberg [50] have presented a framework to manage this collaboration on a shop floor. They identify the challenges of designing "human-robot collaborative (HRC)" systems as defining the health and safety measures (including "collision detection") and programming robots and repeating these steps every time the system is reconfigured. DT's application, in this case, is not much different from other areas; it will offer a safe, virtual environment to conduct tests and validate the system at the design stage. However, it is much more crucial in these cases, considering that testing human-machine systems in real life would expose humans to unpredictable hazards. During operation, DT would again offer real-time monitoring of the physical twin and predictive simulations to identify potential mishaps. They used 3D CAD objects to build the digital twins and imported them into the Tecnomatrix

Process software. These objects (including human models created using the Process Simulate human package) are configured based on the shop floor design and assigned specific tasks. The outcomes of the DT include 1) optimized task allocation for humans and robots so that downtimes and fatigue are as low as possible, 2) efficient shop floor configuration based on the results of four tests, namely "collision analysis" for safety and productivity purposes, "reach test" to find optimum locations, "placement test" a revision of the results of the first two tests, and "vision test" to understand human model POV and remove obstacles, and 3) offline programming and robot visualizations, reducing the testing time in the real world [51, 52].

Malik & Brem [53] present a DT framework for combining the efforts of humans and machines in the workplace. DT could help optimize the shop floor design by ensuring optimized integration, fast reconfiguration, and safety measures. **Figures 5** and **6** show DT's application in an HRC environment and its evolution as the project moves from design to operation and maintenance stages.

Designing such collaborative environments needs to consider the optimum and safest location for robots, equipment, and humans to guarantee access, reduce cycle time, and avoid collisions. DT could provide a virtual environment in which various design scenarios could be tested; moreover, it could help in the development phase when a "mixed reality environment" would allow for the gradual realization of the most optimum design [53].

Malik & Brem [53] also explore the advantages of DT in the operational phase of HRC environments. First, human and robot abilities and skills, as well as their characteristics, must be identified; second, safety measures and their associated limitations are considered; and third, a dynamic task allocation is performed to minimize the idle times of the system. Robot programming is the most challenging and time-consuming part of any robot application; robots can be programmed in the virtual environment (DT). Once their desired operation is achieved, the results

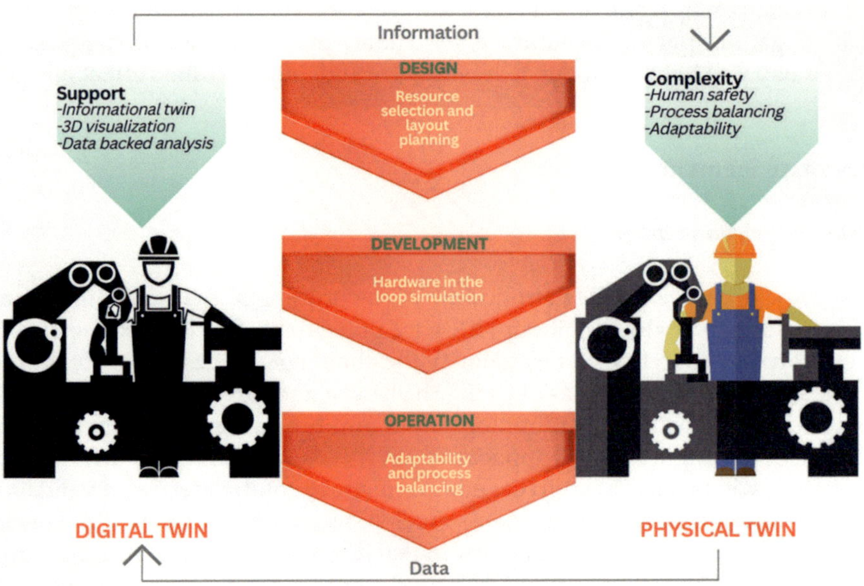

Figure 5.
DT's application in an HRC environment [53].

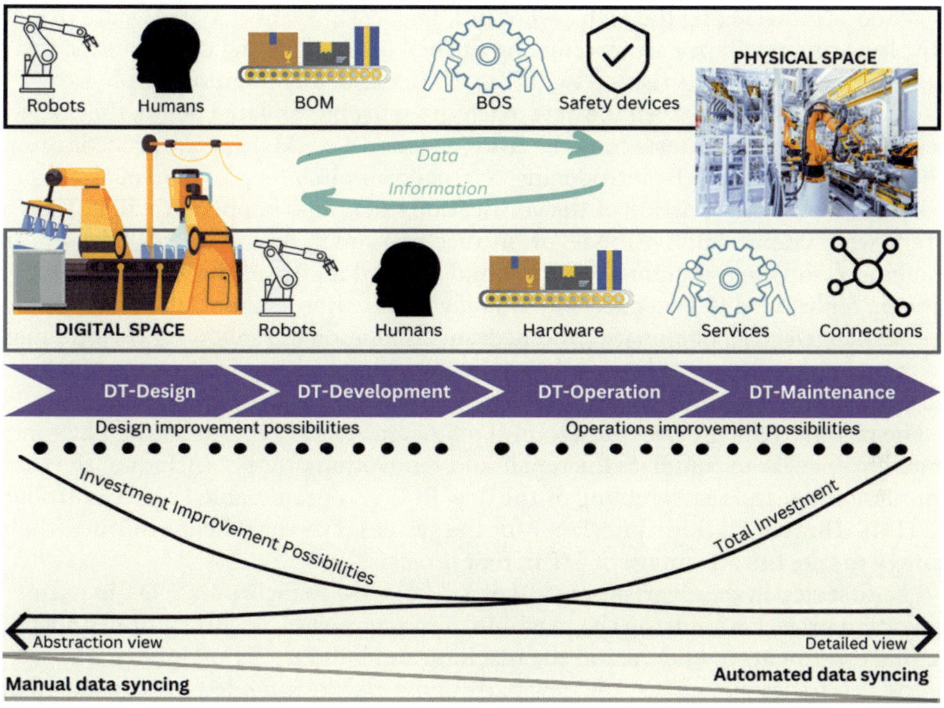

Figure 6.
The evolution of DT as the project evolves [53].

would be transferred to the physical world. This way, real-time simulations could be performed since the digital twins are already designed using robot-understandable language. Like any other implementation of DT, monitoring and performance analysis are also crucial.

2.3.2 Maintenance optimization

Ayani et al. [54] have used DT to reduce the "commissioning time" of machine overhauls or repairs. They argue that as machines age, their repair and renovation become more challenging as the manufacturers stop producing essential parts or technical details. There is also the fact that in the age of Industry 4.0, machines have to support connectivity and data-sharing technologies. Old devices will need several renovations to support such technologies. This idea has great potential in the mining industry as many mines still use old machines in their fleet, and many new ones begin with these machines as they require less capital. But as new mining technologies are introduced, these machines must be updated to be compatible with new, more sophisticated ones and support digital connectivity.

As Ayani et al. [54] point out, machines will be out of operation during repairs, commissioning, and verification, whether they need repair or are just being renovated. Several factors affect this time: machine complexity, resources (time, human, parts), and technical support. Obsolete or missing technical data is the most problematic hindrance, especially if the company conducting the reconditioning is not the manufacturer. Using a DT, a significant portion of the work could be done offline without taking the machines out of operation.

While acknowledging the high cost of building a DT, Ayani et al. [54] state that in the long run, it will pay off by reducing the real commissioning time, energy, and resources required and its risks. Also, DT can be extended to monitor the physical system's whole life cycle. Even the new machines must be validated before they begin operation. These on-site tests could be risky, as failures could delay the procedure or induce unforeseen costs. By introducing "virtual commissioning," Ayani et al. [6] propose that a substantial portion of the verification tests could happen in a digital space if there were a representative model of the machine and its environment. They use an "emulation" software, Simumatik3D, to build a digital replica of the whole system. This software replicates different parts individually, then connects them, and has libraries for essential "electric, pneumatic, and hydraulic" components, allowing the building of a model that reflects its physical counterpart's geometry, physics, and movements. It also supports integrating industrial logic systems to control the digital model.

The reconditioning project at a foundry is Ayani et al.'s [54] case study. They were limited to 4 weeks to complete the repair and renovation process, including the tests, commissioning, and reassembling of the new PLC (Programmable Logic Controller) and HMI (Human Machine Interface) for the system. The chart below outlines their strategy to take full advantage of DT in that project.

The first step in the algorithm Ayani et al. [54] used to incorporate DT in their renovation project is studying the machine, its components, its interaction, and all existing documents to understand the machine and build its digital model. The second stage is to design and test the new parts and software intended to be installed on the system in the virtual space. Then, all these renovations and repairs are integrated into the system, and their functions get tested as a unit during the virtual conditioning. Until now, the machine has been in operation, and there has been little (if any) disturbance. Repairs occur in the last phase, and accurate commissioning and testing happen. Ayani et al. [54] report a shorter time (60%) required for this process section using DT and virtual commissioning.

Guivarch et al. [55] have implemented DT to collect additional data about the loads applied to a helicopter's main rotor. The number of sensors and instrumentation needed for gathering the necessary data mid-operation would be enormous, making DT's application an advantage. Limited sensors would be attached to the rotor to acquire and send data during flight. This information will be inserted into DT and its "multibody simulation" system to predict the loads on all other parts and joints.

Guivarch et al. [55] define DT as a closed loop between the simulations and operational data. They test a laboratory rotor assembly under dynamic and static loadings to train the model. Several sensors (far more than in-flight) are attached to this assembly to acquire the load data. Only the data acquired from the sensors used in flight are the inputs of DT, and other sensors would verify the results and modify the model and sensor positions. The validity of the results depends on the modeling technique and its level of detail.

Okita et al. [56] argue that a change in environment would cause the machine to update its operation and settings accordingly. Therefore, the initial design parameters might not be applicable after a certain amount of service time or if the system experiences an extreme condition that would require drastic changes. They emphasized economic and safety variables as they might be affected by natural disasters, competition, market fluctuations, etc.

The proposed "DTAS" (Digital Twins of Artifact Systems) would enable designers and users of the system to share information during the life cycle of a product or system. This way, they could address these variations and implement necessary system design and functionality changes. A monitoring system (at different scales), a detailed model of the system to reflect the monitoring data, and predictive analysis tools (to test every change before implementing it) are all parts of the DTAS [56].

The benefits of such a system include efficient monitoring plans, identifying weak points of the system in case of a disaster (by conducting simulations), and designing (and testing) mitigation plans in case of an emergency by replicating the disaster in virtual space. Another advantage of using DT is that system builders could avoid over-design and offer flexibility so parts of the system or its operation could be changed if necessary (DT would monitor and predict if there is a need for change) [56].

"DTAS" helps obtain up-to-date technical information and validate new PLC and HMI. One of the most essential capabilities of the system is providing a shared, unified information hub for all parties involved. All parties will have their own needs and priorities, which requires different types of models to be included in the DT (or one unified model) [56].

To investigate the impacts of economic factors on the mining industry's maintenance efforts, Savolainen & Urbani [57] use a combination of Discrete Event Simulation (DES) for maintenance and System Dynamics (SD) for "managerial cash flow (CF)" to "co-simulate" maintenance decisions and their financial implications. This "co-simulation" without DT would not be possible as they are fundamentally different and require data transfer.

The economic variable in the Savolainen & Urbani [57] model is the monthly metal price and the up-to-minute maintenance model of the mine fleet, which determine the operation's profitability. The goal is to use DT as an optimization tool (decision-making based on a set of heuristic rules) for maintenance activities. The maintenance for one component depends on "time to failure" (TTF) distribution and the cost of corrective maintenance. But simultaneously, a multi-component system could be managed as one unit (since components are interdependent).

There are two types of maintenance strategies, "condition-based (CBM)" and "time-based (TBM)," and both include two activities: "corrective" (CM) and "preventive" (CM), which is justified by potentially lower costs than the failure of the system. DT would also include simulation tools to estimate the machine's failure time and suggest preventive maintenance [57].

Savolainen & Urbani [57] divide the mine systems into "server-queue" (shovels, dump sites, and workshop) and "agents" (trucks) that are free to move between the "FIFO servers" (*first in, first out*). In addition to being a server, shovels are subject to aging and failure and could unexpectedly fail; in this case, they are transferred to the workshop with the highest priority. Each time a truck completes an event (between two servers), the model decides whether to send it to the workshop or another server (not the workshop). As for the shovels, this inspection happens after each truck loading to determine whether they need to be sent to preventive maintenance. Ultimately, the maintenance policy would depend on the allowable time for trucks (agents) and shovels to work without any preventive maintenance. Determining these time thresholds (a function of scale and mean failure intervals) depends on minimizing the cost function associated with the operation (maintenance and failure) while satisfying a bounding condition, the minimum production target.

The results of the maintenance module are transferred into a CF model that also includes metal price (simulated using a specific economic model to represent market uncertainty). While it is possible to put both simulations into one software, there are practical challenges. Mine managers are hesitant to share sensitive data, and it's expensive to develop this kind of software [57].

To examine the conformity of the model to the logical trade-offs between PM and CM costs, Savolainen & Urbani [57] also conducted a sensitivity analysis focused on the available resources (trucks, shovels, and workshops) which yield complicated results because different configurations might yield the exact cost. However, as Savolainen & Urbani [57] point out, several other aspects are worth considering. Chief among them are the capital investment requirement, required human resources, and the integrity of the assembled system.

Savolainen & Urbani [57] conducted two experiments to examine the whole DT approach's validity. The first one is a fleet design based on three different maintenance policies ("max preventive," "max corrective," and "balanced"). The results show that the highest profits and mill utilization are acquired for different fleet configurations in case of a "run-to-failure" maintenance policy. The optimal number of trucks and shovels is 8–9 trucks and 1–2 shovels, resulting in 90% + mill utilization. In the next step, they used cost distribution to reflect the uncertainties related to the costs of CM and PM (5 and 1.1 times the assumed value, respectively) and rerun the simulations. The results are similar to the first version: no PM, nine trucks, one shovel, and 98% mill utilization. The writers argue that this is due to the limited capacity of workshops. But logically, the best maintenance policy has to find the optimal maintenance intervals that would keep the number of CMs to a minimum.

The other experiment conducted by Savolainen & Urbani [57] is to find the optimal PM intervals by running the simulations for different numbers of OM events per week of simulations. The range is from 1 PM event every 8 weeks to 1 PM event per week. The results indicate that nine trucks, one shovel, and 1 PM per week would yield the highest profits with 91.6% mill utilization. However, a closer examination of the results indicates that managerial decision-making must be implemented as there is another option with fewer trucks and slightly longer PM intervals that would yield profits in the range of the most profitable option.

Another important finding of this optimization simulation is the optimal number of trucks per shovel based on the workshops available and PM intervals. DT helps combine a DES optimization and a CF model to find the best fleet design and maintenance policy for each mine based on the available resources and the mineral value. However, issues must be addressed, including the scalability of complex models and data, data integration and transfer between the modeling disciplines, and verifying the results [57].

2.3.3 Autonomous vehicles (AVs)

As for autonomous vehicles, the existing literature has a positive outlook toward applying DT technology in their control and maintenance. Almeaibed et al. [58] mention the applicability of DT in data transformation, data processing, and analysis of AVs (defined as vehicles that can sense all their surroundings and move without human intervention) and intelligent manufacturing systems. But they also acknowledge the vulnerability of such systems to malicious attacks and propose a protective method to assure data security within the digital twins' paradigm (**Figure** 7).

The argument is that if a high-fidelity virtual model represents the physical system's past, current, and future states, it would be plausible to identify data anomalies by the apparent deviation of the virtual and real-world objects. According to the researchers, the five steps of AV design include "perception"

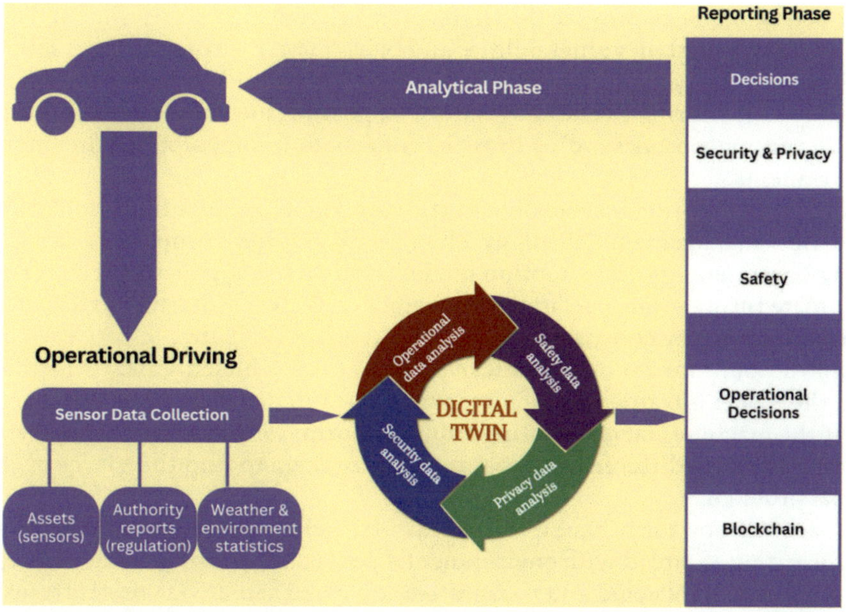

Figure 7.
DT supports autonomous vehicles' safety and security [58].

(environmental sensibility using LiDARs and Radars), "localization" (the exact coordinates of the vehicle), "planning" (based on the first two steps, what is the course of action), "control" (execution of the "plans"), and "system management" (an HMI to provide the user with necessary information such as malfunctions within the system) [58].

These steps must be thoroughly examined, and DT could provide a safe environment for such examinations. When describing the elements of the DT for AVs, they mention the initial state in which the DT has all the manufacturing data; when the system becomes "operational," the sensors begin collecting data and feeding the digital twins, which examines the soundness of the information received and analyzes them, then either the DT makes an automatic decision (highest level of autonomy) or reports the results back for decision-making. The biggest challenge in realizing this automatic response is creating a system capable of receiving vast amounts of data (high levels of depth and detail) in real time (or near real time), analyzing it, and reporting the results [58].

In the cybersecurity issue of AVs, the biggest challenge is the enormity of the system, which comprises many sensors and connection networks, and each one of those sensors could be a potential attack point. Dependency on cloud computing, the use of diverse open-source programming languages without validation, and the growing hesitation of manufacturers to share resources and technologies are among the notable focal points when it comes to the cybersecurity of AVs [58].

As examples of the possible attacks, Almeaibed et al. [58] point out attacks on the individual sensors, internal (passwords and keys), and external networks (a "vehicle to everything" attack that affects any system providing connection to the cloud which could lead to "distributed denial of service" attacks), and "spoofing attacks" on GPS for instance that would give the attacker the ability to manipulate the vehicle by sending erroneous data instead of accurate GPS information. DT's role in this section would be to provide the testbed for different attack scenarios and failure modes to identify system vulnerabilities and behavior in case of malicious actions. Understanding the system's shortcomings would help provide security and safety countermeasures during the design phase. One possible identification method is to detect one or two sensors sending irregular information compared to other monitoring instruments.

Energy consumption, especially electric vehicles, is another important focal point in fleet management. Zhaolong Zhang et al. [59] have proposed a machine learning-based "energy consumption model" trained using real-world data and can simulate future events. As multiple correlated factors (internal, external, and human) affect energy consumption, DT would offer the chance to take each of those into account for a more holistic overview of the vehicle's state. As mentioned earlier, the model is trained using historical data, and expected behavior is set in place; if the machine starts deviating from the norm, for instance, that will trigger the system to check different sensors and analyze them to find the source of the potential problem.

As machines nowadays have multiple sensors reading the vehicle's current state, this information, coupled with environmental data (temperature, wind, traffic, road condition, etc.), is collected and transmitted to a cloud server via wireless transmission (5G). The aggregated data would be analyzed and compared to historical data to update the digital model and identify anomalies to make necessary adjustments to the physical twin. All the analyses, simulations, and results are visualized using a semantic 3D model to make decision-making much more manageable. Each time the model makes a prediction, the results are compared to the real world to examine the accuracy of the simulations, and the results of such comparisons are used to evolve and optimize the model [59].

Zhaolong Zhang et al.'s study [59] indicates that data obtained from the vehicle encompassed parameters such as "state of charge (SOC)," "instantaneous output voltage and current of the battery," and battery temperature range. The readings happened at a constant temperature and 80 "instances" to define a "compensation coefficient," which indicates the difference between the physical and digital twins. They could attain the relationship between the two by calculating the compensation coefficient at different temperatures. By knowing that correlation, it would be possible to simulate the vehicle's future state and estimate the system's errors. While they acknowledge that other factors affect the measurements and errors (such as wind direction, slopes, road conditions, etc.), this is an example of the correcting process necessary to optimize the model.

Some scenarios would require a subjective study of crucial parts of a vehicle. For example, monitoring autonomous vehicles' braking systems is a significant health and safety measure. Magargle et al. [60] have used DT for heat monitoring and predictive maintenance of an automotive braking system. They use physics-based models with assigned failure modes and train them (using ML algorithms) to simulate the future state of "degradable" components. By using reduced-order models for parts of the braking system, it would be possible to detect the effect of each element. By comparing it to regular rates, Magargle et al. [60] have defined a "wear rate" to predict maintenance time and to account for faulty conditions (like when the ABS is disconnected). DT, in this case (like many others), is an integrated space that enables receiving data from the asset, monitoring, and controlling it using the inherent HMI.

Another potential application for DT is the health monitoring of machines and equipment and their specific parts of interest. Tuegel et al. [61] have used DT to study the structural evolution of aircraft in its life cycle. They introduce a vision (concept) that, if realized, would make it possible to build an "ultrahigh-fidelity model" that would generate 1 petabyte of data in a one-hour simulated flight, and if this simulation is repeated for 1000 hrs. The amount of data for predictive analysis and finding the "unpredicted undesirables" would be enormous. Remember that all these analyses happen before the real aircraft's first real-life flight. When the aircraft becomes operational, the same model is used for monitoring, risk analysis, and controlling its physical twin.

Though the modeling and computation limitations only allow for a portion of Tuegel et al.'s [61] concept to become operational, it is still an enormous progress. One important issue, like any other simulation, is that there will be errors and simplifications that would affect the results. Still, as DT is a dynamic system, by conducting tests and training it (using data-driven models and ML algorithms), it would be able to quantify the errors and give a reliable operation prediction.

The direct relationship between the twins as one receives the data, stores it, processes it, conducts studies, and finally presents the data in the form of information back to the other, as well as storing each step of the structural evolution of the system and its components and updating itself (thus avoiding worst-case scenario simulations that leads to over-design), is the main advantage of DT technology in machines health monitoring. DT also offers the opportunity to use various modeling methods and data types for each part of the same machine and integrates all of them in a unified space (**Figure 8**) [61].

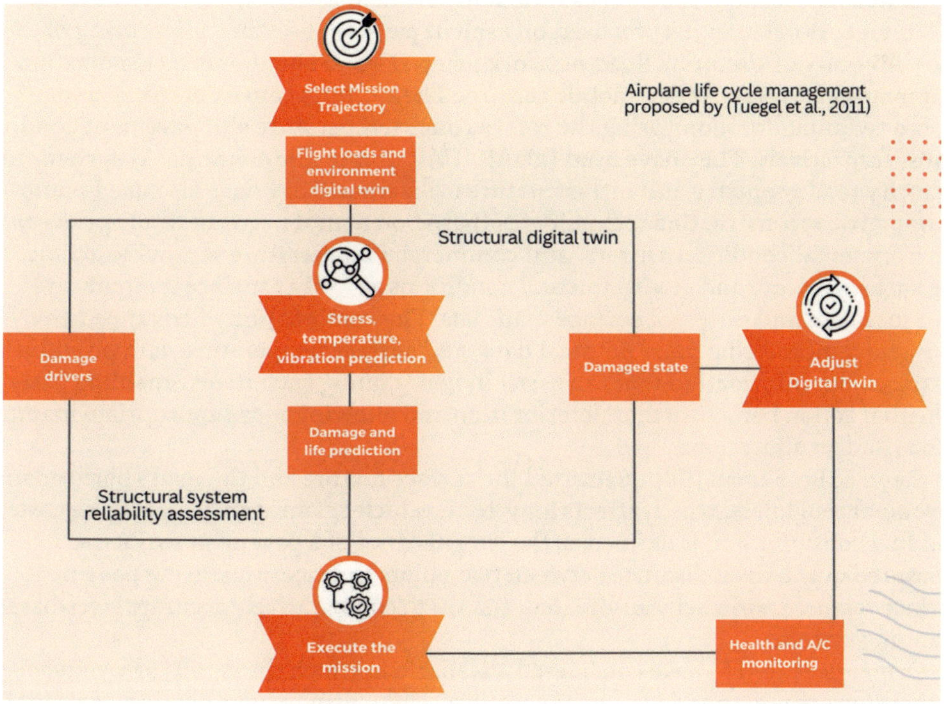

Figure 8.
Airplane life cycle management schematic using DT [61].

2.3.4 Road maintenance

Road design and maintenance are vital operations at the mine site; the slightest problems would cause severe damage to giant machines carrying heavy loads and disrupt their optimum operation [62].

Machl et al. [63] argue that agricultural machines are becoming heavier and faster, and the distance they must travel due to the increasing size of the farms is also increasing. This development has many similarities to the mine sites. To identify the existing roads to modernize them and design a needs-based core road network, they use a DT of the cultivated landscape, which contains the necessary spatiotemporal data and analytical tools that enable monitoring and assessment of the impact of each design scenario. The DT is designed to perform state-wide monitoring and analysis of all farm-field transportation roads and their relations. In this case, the main challenge is the distributed and heterogeneous data across stakeholders.

Compared to conventional landscape models, DT requires a broader scope of information that includes semantic information about topographic objects, like agricultural parcels, their crop rotation, farms, and the cultivated area connected to a specific road segment and the evolution of the landscape over time. Semantic enrichment allows assigning additional information to individual objects, facilitating an integrated view of the cultivated landscape as a complex system. The source of semantic data could be external information systems, real-time sensor readings, or the output of complex analytical methods. Machl et al. [63] have introduced methods for estimating the agricultural transportation paths, describing the geometry of agricultural parcels, and detecting, analyzing, and documenting the changes in the cultivated landscape at the level of individual elements.

Steyn & Broekman [64] focused on asphalt pavement surface monitoring of the University of Pretoria's Road network using DT and simultaneous localization and mapping by optical and mobile sensors. Their efforts consist of micro and macro twinning for monitoring the road surface temperature and pavement conditions, respectively. They have used LiDAR, UAV, and traffic counting AI systems to quantify road geometry and infrastructure utilization. They have also used photogrammetric reconstruction technologies (based on neural networks), proprietary environmental condition sensors, and commercial temperature sensors to acquire the surface texture and environmental conditions as well as surface temperature at four locations and different surface materials. The combination of advanced environmental monitoring data, physical data, and surface temperature data provides management information that can assist in maintaining such roads, enabling early remedial action before further deterioration and allowing more appropriate maintenance budget allocations.

Steyn & Broekman [64] quantified the surface texture and the road's performance. Pavement roughness, which affects long-term vehicle maintenance, operating costs, and fuel consumption, is defined as the irregularities of a pavement surface as measured over a fixed distance between two points in space. Measuring pavement surface texture is primarily automated, thanks to comparatively accurate laser-based scanning systems.

Ormándi et al. [65] have considered a scenario in which the details of a vehicle's suspension are not provided or not accessible. Then, using a combination of virtual and real sensors to compare the real-life and simulated measurements iteratively, they came up with an estimation for the suspension characteristics. A car model is built using IPG Carmaker based on the software's default settings. Then, a "meta-heuristic optimization tool such as genetic algorithm (GA)" is used to compare the real-life sensor readings with the virtual measurements (available in the IPG environment) based on a defined cost function. The measurements include "vertical acceleration," velocity of the car, and the location coordinates when traveling over a speed bump. Based on the GA analysis results, the virtual model's suspension parameters get updated until an acceptable threshold is reached.

For further validations, Ormándi et al. [65] used OpenCRG to create a road surface with great details (scanned by lasers) and import it to the IPG carmaker's environment. The results show acceptable agreement with real-life measurements. However, since the IPG carmaker is incapable of real-time synchronization of the car and its model (the exact position of the car cannot be set), they have used Unity3D for another round of tests. Importing the OpenCRG files to the Unity3D environment is challenging but achievable. Here, the real-time connection between the car and its model is achieved based on the location readings. Building upon the calculated suspension parameters from GA, the simulations in the Unity3D environment are also comparable to the accurate sensor readings.

3. Conclusion

The Digital Twins (DT) approach holds immense promise for the mining industry. It enables real-time monitoring and modeling of various mining processes, from geotechnical aspects to drilling and blasting, fleet management, and energy efficiency. This technology empowers mining operators to optimize production, reduce delays, and enhance worker safety (**Figures 9** and **10**).

Root cause analysis facilitated by DT's data-driven models provides a powerful tool for identifying and rectifying operational shortcomings. However, implementing DT in mining faces challenges like data management, cybersecurity, integration, communication, and personnel training. Successful adoption will require collaboration among mining companies, technology providers, and stakeholders.

Ongoing research must focus on refining the DT framework, improving data acquisition and storage methods, and enhancing joint extraction algorithms. Additionally, advancements in surface temperature and moisture content acquisition systems using RGB images will benefit bench characterization and slope stability studies.

The ideal candidate for DT addresses specific business challenges or explores new opportunities. Identifying one involves improving asset performance, reducing downtime, and increasing production or throughput. Assessing DT readiness requires ranking business impact against technical feasibility, considering factors like infrastructure connectivity, data access, appetite for change, and organizational maturity. Technical readiness is assessed through factors like Operational Technology (OT) complexity, IT complexity, analytics, systems complexity, and project readiness. High business impact and technical readiness are prerequisites for assessing the Minimum Viable Product (MVP). Implementing the DT methodology requires technological capabilities, talent, collaborations, partnerships, alliances, and an organizational structure conducive to agility and innovation. Tools like The Oracle IOT DT framework, Azure DT, IBM DT Exchange, Ansys Twin Builder, and GE's Predix Platform are utilized for building DTs. The development process must be agile and structured

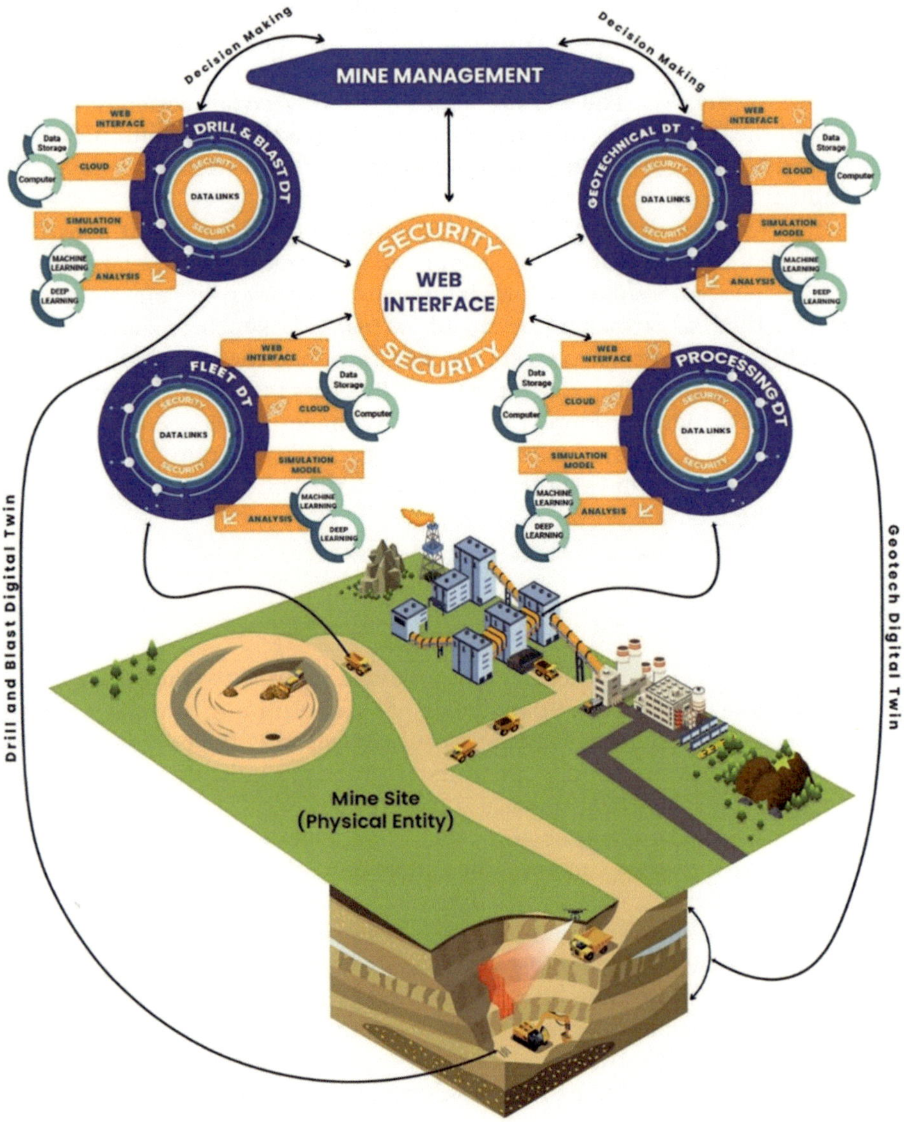

Figure 9.
Schematic of the proposed mining DT.

and consider technical and business impacts. Key considerations include defining problems addressed by DT, identifying beneficiaries, stating the unique value proposition (UVP), determining top capabilities, addressing external challenges, creating a business case, specifying key metrics, planning for integration, and assessing costs [4].

Ultimately, the potential of Digital Twins technology to revolutionize mining, improve efficiency, reduce costs, and minimize environmental impact cannot be understated. By embracing DT, mining companies can bolster competitiveness and embrace a sustainable future. Continued research and development efforts are essential to unlock the digital twins' full potential in the mining industry.

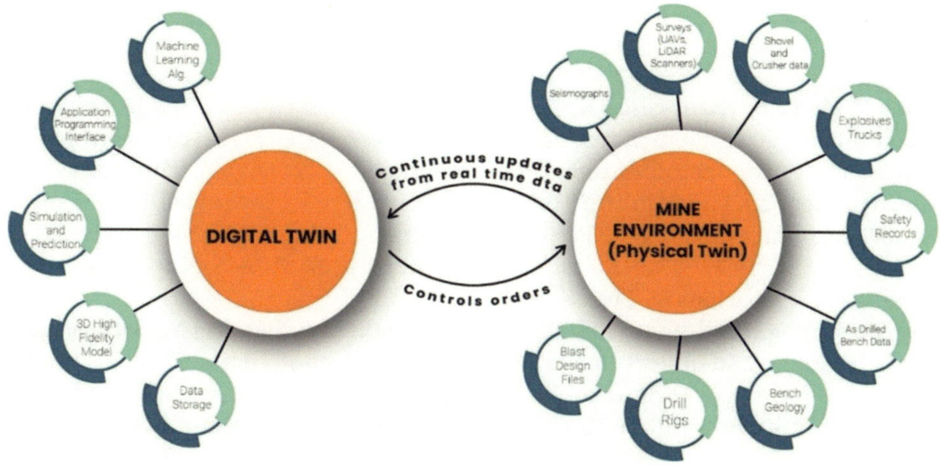

Figure 10.
Details of the relationship between physical and digital twins (D&B DT).

Author details

Milad Ghahramanieisalou* and Javad Sattarvand
University of Nevada, Reno, NV, USA

*Address all correspondence to: milad.gh@nevada.unr.edu

References

[1] Jiang F, Ma L, Broyd T, Chen K. Digital twin and its implementations in the civil engineering sector. Automation in Construction. 2021;**130**(July):103838. DOI: 10.1016/j.autcon.2021.103838

[2] Zheng X, Lu J, Kiritsis D. The emergence of cognitive digital twin: Vision, challenges and opportunities. International Journal of Production Research. 2022;**60**(24):7610-7632. DOI: 10.1080/00207543.2021.2014591

[3] Negri E, Fumagalli L, Macchi M. A review of the roles of digital twin in CPS-based production systems. Procedia Manufacturing. 2017;**11**(June):939-948. DOI: 10.1016/j.promfg.2017.07.198

[4] Nath SV, Schalkwyk P, v., Isaacs, D. Building Industrial Digital Twins: Design, Develop, and Deploy Digital Twin Solutions for Real-World Industries Using Azure Digital Twins. Packt Publishing; 2021

[5] Söderberg R, Wärmefjord K, Carlson JS, Lindkvist L. Toward a digital twin for real-time geometry assurance in individualized production. CIRP Annals - Manufacturing Technology. 2017;**66**(1):137-140. DOI: 10.1016/j.cirp.2017.04.038

[6] Duan JG, Ma TY, Zhang QL, Liu Z, Qin JY. Design and application of digital twin system for the blade-rotor test rig. Journal of Intelligent Manufacturing. 2023;**34**(2):753-769. DOI: 10.1007/S10845-021-01824-W

[7] Kor M, Yitmen I, Alizadehsalehi S. An investigation for integration of deep learning and digital twins towards construction 4.0. Smart and Sustainable Built Environment. 2023;**12**(3):461-487. DOI: 10.1108/SASBE-08-2021-0148

[8] Ma C, Gui H, Liu J. Self learning-empowered thermal error control method of precision machine tools based on digital twin. Journal of Intelligent Manufacturing. 2023;**34**(2):695-717. DOI: 10.1007/S10845-021-01821-Z

[9] Ghahramanieisalou M, Sattarvand J. Applications of Digital Twin Technology in Productivity Optimization of Mining Operations. In: APCOM. Rapid City, SD, USA: SME; 2023. pp. 1-17

[10] Alizadehsalehi S, Yitmen I. Digital twin-based progress monitoring management model through reality capture to extended reality technologies (DRX). Smart and Sustainable Built Environment. 2023;**12**(1):200-236. DOI: 10.1108/SASBE-01-2021-0016/FULL/HTML

[11] Ramu SP, Boopalan P, Pham QV, Maddikunta PKR, Huynh-The T, Alazab M, et al. Federated learning enabled digital twins for smart cities: Concepts, recent advances, and future directions. Sustainable Cities and Society. 2022;**79**:103663. DOI: 10.1016/J.SCS.2021.103663

[12] Perno M, Hvam L, Haug A. Implementation of digital twins in the process industry: A systematic literature review of enablers and barriers. Computers in Industry. 2022;**134**:103558. DOI: 10.1016/J.COMPIND.2021.103558

[13] Agrawal A, Fischer M, Singh V. Digital twin: From concept to practice. Journal of Management in Engineering. 2019;**38**(3):06022001

[14] Li L, Lei B, Mao C. Digital twin in smart manufacturing. Journal of Industrial Information Integration. 2022;**26**:100289. DOI: 10.1016/J.JII.2021.100289

[15] Autiosalo J, Ala-Laurinaho R, Mattila J, Valtonen M, Peltoranta V, Tammi K. Towards integrated digital twins for industrial products: Case study on an overhead crane. Applied Sciences. 2021;**11**:683. DOI: 10.3390/app11020683

[16] Botín-Sanabria DM, Mihaita S, Peimbert-García RE, Ramírez-Moreno MA, Ramírez-Mendoza RA, Lozoya-Santos J d. Digital twin technology challenges and applications: A comprehensive review. Remote Sensing. 2022;**14**:1335. DOI: 10.3390/RS14061335

[17] Grieves M, Vickers J. Digital twin: Mitigating unpredictable, undesirable emergent behavior in complex systems. In: Kahlen J, Flumerfelt S, Alves A, editors. Transdisciplinary Perspectives on Complex Systems. Cham: Springer; 2017. DOI: 10.1007/978-3-319-38756-7_4

[18] Lee J, Lapira E, Yang S, Kao A. Predictive manufacturing system-trends of next-generation production systems. IFAC Proceedings Volumes. 2013;**46**(46):150-156. DOI: 10.3182/20130522-3-BR-4036.00107

[19] Schluse M, Rossmann J. From simulation to experimentable digital twins: Simulation-based development and operation of complex technical systems. In: 2016 IEEE International Symposium on Systems Engineering (ISSE). Edinburgh, UK; 2016. pp. 1-6. DOI: 10.1109/SysEng.2016.7753162

[20] Gabor T, Belzner L, Kiermeier M, Beck MT, Neitz A. A simulation-based architecture for smart cyber-physical systems. In: 2016 IEEE International Conference on Autonomic Computing (ICAC), Wuerzburg, Germany. 2016. pp. 374-379. DOI: 10.1109/ICAC.2016.29

[21] Kraft EM. The US air force digital thread/digital twin – Life cycle integration and use of computational and experimental knowledge. In: 54th AIAA Aerospace Sciences Meeting. AIAA 2016-0897. Jan 2016. pp. 1-22. DOI: 10.2514/6.2016-0897

[22] Rasheed A, San O, Kvamsdal T. Digital twin: Values, challenges and enablers from a modeling perspective. IEEE Access. 2020;**8**:21980-22012. DOI: 10.1109/ACCESS.2020.2970143

[23] Kamble SS, Gunasekaran A, Parekh H, Mani V, Belhadi A, Sharma R. Digital twin for sustainable manufacturing supply chains: Current trends, future perspectives, and an implementation framework. Technological Forecasting and Social Change. 2022;**176**:121448. DOI: 10.1016/J.TECHFORE.2021.121448

[24] Kapteyn MG, Knezevic DJ, Huynh DBP, Tran M, Willcox KE. Data-driven physics-based digital twins via a library of component-based reduced-order models. International Journal for Numerical Methods in Engineering. 2019;**2020**:1-18. DOI: 10.1002/nme.6423

[25] Kapteyn M, Willcox KE. Predictive digital twins: Where dynamic data-driven learning meets physics-based modeling. In: Darema F, Blasch E, Ravela S, Aved A, editors. Dynamic Data Driven Applications Systems. DDDAS 2020. Lecture Notes in Computer Science. Vol. 12312. Cham: Springer; 2020. pp. 1-10. DOI: 10.1007/978-3-030-61725-7_1

[26] Wang Y, Xu R, Zhou C, Kang X, Chen Z. Digital twin and cloud-side-end collaboration for intelligent battery management system. Journal of Manufacturing Systems. 2022;**62**:124-134. DOI: 10.1016/J.JMSY.2021.11.006

[27] Kane GC, Palmer D, Philips Nguyen A, Kiron D, Buckley N. Strategy, Not Technology, Drives

Digital Transformation. No. 57181. MIT Sloan Management Review & Deloitte; 14 Jul 2015. Available from: http://sloanreview.mit.edu/projects/strategy-drives-digital-transformation/

[28] Grieves MW. Virtually intelligent product systems: Digital and physical twins. Complex Systems Engineering: Theory and Practice. 2019;**256**:175-200

[29] Grieves M. Digital Twin: Manufacturing Excellence through Virtual Factory Replication. White Paper. 2014;**1**:1-7

[30] Read J, Stacey P. Guidelines for Open Pit Slope Design. Clayton: SCIRO Publishing; 2010. DOI: 10.1071/9780643101104

[31] Choi SH, Park KB, Roh DH, Lee JY, Mohammed M, Ghasemi Y, et al. An integrated mixed reality system for safety-aware human-robot collaboration using deep learning and digital twin generation. Robotics and Computer-Integrated Manufacturing. 2022;**73**:102258. DOI: 10.1016/J.RCIM.2021.102258

[32] Aheleroff S, Xu X, Zhong RY, Lu Y. Digital twin as a service (DTaaS) in industry 4.0: An architecture reference model. Advanced Engineering Informatics. 2021;**47**(August 2020):101225. DOI: 10.1016/j.aei.2020.101225

[33] Zhao L, Bi Z, Hawbani A, Yu K, Zhang Y, Guizani M. ELITE: An intelligent digital twin-based hierarchical routing scheme for Softwarized vehicular networks. IEEE Transactions on Mobile Computing. 1 Sep 2023;**22**(9):5231-5247. DOI: 10.1109/TMC.2022.3179254

[34] Norris JE, Stokes A, Mickovski SB, Cammeraat E, van Beek R, Nicoll BC, et al. Slope Stability and Erosion Control: Ecotechnological Solutions. Dordrecht: Springer; 2008. DOI: 10.1007/978-1-4020-6676-4

[35] Wyllie DC. Rock Fall Engineering. 1st ed. CRC Press; 2015. DOI: 10.1201/b17470

[36] Battulwar R. Automatic Extraction of Joint Characteristics from Rock Mass Surface Point Cloud Using Deep Learning. Reno: University of Nevada; 2021

[37] Peik B. A New Three-Dimensional Rockfall Trajectory Simulator for Open-Pit Mines. Reno: University of Nevada; 2020

[38] Fan C, Zhang C, Yahja A, Mostafavi A. Disaster City digital twin: A vision for integrating artificial and hu-man intelligence for disaster management. International Journal of Information Management. 2021;**56**(December 2019):102049. DOI: 10.1016/j.ijinfomgt.2019.102049

[39] Losier LM, Fernandes R, Tabarro P, Braunschweig F. The Importance of Digital Twins for Resilient Infrastructure. Vol. 102019. Available from: https://cdn2.webdamdb.com/md_A6HafPVAhHf0.jpg.pdf

[40] Kamari M, Ham Y. AI-based risk assessment for construction site disaster preparedness through deep learning-based digital twinning. Automation in Construction. 2022;**134**(July 2021):104091. DOI: 10.1016/j.autcon.2021.104091

[41] Joshi DR, Eustes AW III, Rostami J. Testing lunar regolith characterization algorithms with simulated subsurface samples and digital twin data. In: 52nd Lunar and Planetary Science Conference. 2021 (LPI Contrib. No. 2548)

[42] Gomez Llerena JA, Ghahramanieisalou M, Sattarvand J.

Tackling geotechnical risks in tailings dams using high-resolution UAV imaging and advanced image processing. In: Geo-Risk 2023: Innovation in Data and Analysis Methods, Arlington, Virginia, USA. ASCE; 2023. pp. 220-228. DOI: 10.1061/9780784484975.024

[43] Lu R, Brilakis I. Digital twinning of existing reinforced concrete bridges from labelled point clusters. Automation in Construction. 2019;**105**(February):102837. DOI: 10.1016/j.autcon.2019.102837

[44] Shim CS, Dang NS, Lon S, Jeon CH. Development of a bridge maintenance system for prestressed concrete bridges using 3D digital twin model. Structure and Infrastructure Engineering. 2019;**15**(10):1319-1332. DOI: 10.1080/15732479.2019.1620789

[45] Zhang H, Wang R, Wang C. Monitoring and warning for digital twin-driven mountain geological disaster. In: 2019 IEEE International Conference on Mechatronics and Automation (ICMA), Tianjin, China. 2019. pp. 502-507. DOI: 10.1109/ICMA.2019.8816292

[46] Ford DN, Wolf CM. Smart cities with digital twin Systems for Disaster Management. Journal of Management in Engineering. 2020;**36**(4):04020027. DOI: 10.1061/(asce)me.1943-5479.0000779

[47] Gokhale BV. Rotary Drilling and Blasting in Large Surface Mines. 1st ed. CRC Press; 2010. DOI: 10.1201/b10972

[48] Steir JF. ISEE Blasters' Handbook. 18th ed. International Society of Explosives Engineers (ISEE); 2011. Available from: https://books.google.com/books?id=Jr-gmgEACAAJ

[49] Hustrulid WA. Blasting Principles for Open Pit Mining. Vol. 1. A. A. Balkema; 1999. Available from: https://books.google.com/books?id=bc2YXfXvmh4C

[50] Malik AA, Bilberg A. Digital twins of human robot collaboration in a production setting. Procedia Manufacturing. 2018;**17**:278-285. DOI: 10.1016/j.promfg.2018.10.047

[51] Kousi N, Gkournelos C, Aivaliotis S, Lotsaris K, Bavelos AC, Baris P, et al. Digital twin for designing and reconfiguring human-robot collaborative assembly lines. Applied Sciences. 2021;**11**:4620. DOI: 10.3390/app11104620

[52] Stączek P, Pizoń J, Danilczuk W, Gola A. A digital twin approach for the improvement of an autonomous mobile robots (AMR's) operating environment—A case study. Sensors. 2021;**21**:7830. DOI: 10.3390/s21237830

[53] Malik AA and Brem A. Man, Machine and Work in a Digital Twin Setup: A Case Study. 2020. Available from: http://arxiv.org/abs/2006.08760

[54] Ayani M, Ganebäck M, Ng AHC. Digital twin: Applying emulation for machine reconditioning. Procedia CIRP. 2018;**72**:243-248. DOI: 10.1016/j.procir.2018.03.139

[55] Guivarch D, Mermoz E, Marino Y, Sartor M. Creation of helicopter dynamic systems digital twin using multi-body simulations. CIRP Annals. 2019;**68**(1):133-136. DOI: 10.1016/j.cirp.2019.04.041

[56] Okita T, Kawabata T, Murayama H, Nishino N, Aichi M. A new concept of digital twin of artifact systems: Synthesizing monitoring/inspections, physical/numerical models, and social system models. Procedia CIRP. 2019;**79**:667-672. DOI: 10.1016/j.procir.2019.02.048

[57] Savolainen J, Urbani M. Maintenance optimization for a multi-unit system with digital twin simulation: Example from the mining industry. Journal of Intelligent Manufacturing. 2021;**32**(7):1953-1973. DOI: 10.1007/s10845-021-01740-z

[58] Almeaibed S, Al-Rubaye S, Tsourdos A, Avdelidis NP. Digital twin analysis to promote safety and security in autonomous vehicles. IEEE Communications Standards Magazine. 2021;**5**(1):40-46. DOI: 10.1109/MCOMSTD.011.2100004

[59] Zhang Z, Zou Y, Zhou T, Zhang X, Xu Z. Energy consumption prediction of electric vehicles based on digital twin technology. World Electric Vehicle Journal. 2021;**12**(4):1-13. DOI: 10.3390/wevj12040160

[60] Magargle R, Johnson L, Mandloi P, Davoudabadi P, Kesarkar O, Krishnaswamy S, et al. A simulation-based digital twin for model-driven health monitoring and predictive maintenance of an automotive braking system. In: Proceedings of the 12th International Modelica Conference, Prague, Czech Republic. 2017. 15-17 May 2017. DOI: 10.3384/ecp1713235

[61] Tuegel EJ, Ingraffea AR, Eason TG, Spottswood SM. Reengineering aircraft structural life prediction using a digital twin. International Journal of Aerospace Engineering. 2011;**2011**:14. Article ID 154798. DOI: 10.1155/2011/154798

[62] Marais WJ, Thompson RJ, Visser AT. Managing mine road maintenance interventions using mine truck on-board data. In: The International Conference on Surface Mining 2008 'Challenges, Technology, Systems and Solutions'. Johannesburg, South Africa: The Southern African Institute of Mining and Metallurgy; 2008. pp. 3-14

[63] Machl T, Donaubauer A, Kolbe TH. Planning agricultural core road networks based on a digital twin of the cultivated landscape. Journal of Digital Landscape Architecture. 2019;**2019**(4):316-327. DOI: 10.14627/537663034

[64] Steyn WJ, Broekman A. Development of a digital twin of a local road network: A case study. Journal of Testing and Evaluation. 2021;**51**(1):20210043. DOI: 10.1520/jte20210043

[65] Ormándi T, Varga B, Tettamanti T. Estimating vehicle suspension characteristics for digital twin creation with genetic algorithm. Periodica Polytechnica Transportation Engineering. 2021;**49**(3):231-241. DOI: 10.3311/PPTR.18576

Chapter 3

Evolution of a Technosol Cultivated with Perennial Grass, over 15 Years: Potential Use as Carbon Sinks

Luís Eduardo Torma Burgueño, Luiz Fernando Spinelli Pinto, Lizete Stumpf, Clause Fátima de Brum Piana, Maurício Silva de Oliveira, Maurizio Silveira Quadro and Pablo Miguel

Abstract

In this study, we investigated the increase in carbon in the surface layer of a technosol, promoted by perennial grasses, and its implications for the recovery of land degraded by coal mining in southern Brazil, and its potential as carbon sinks. To monitor the recovery process of soils constructed after coal mining, a randomized block experiment was implemented in 2003, with several species of perennial grasses, in a coal mine in southern Brazil. Over 15 years of monitoring, the species *Hemarthria altissima* and *Urochloa brizantha* showed the best rates of incorporation of organic matter and soil decompaction capacity. To evaluate the efficiency of these species over time in recovering these technosols, polynomial regression models were adjusted to total organic carbon (TOC) and soil bulk density (Bd) data. The ability of these species to increase organic soil carbon stocks (SCS) was also estimated. The results verified showed that the use of perennial grasses in the recovery processes of these areas can, over time, promote an increase in TOC (425%) and reduce Bd. In this sense, it is believed that technosols constructed in mining areas can become excellent carbon sinks if properly designed for this purpose.

Keywords: technosols, greenhouse gases reduction, recovery of degraded areas, coal mining, polynomial regression

1. Introduction

In Brazil, coal reserves are concentrated in the State of Rio Grande do Sul (RS), which holds approximately 80% of the country's coal reserves. In the municipality of Candiota, in the Campanha region, there is the largest surface coal mine, operated by Companhia Riograndense de Mineração (CRM) since 1961, with around one billion tons, at depths of up to 50 m. Although coal mining for energy production can be considered a strategic resource for national development [1], it is also one of the most harmful anthropogenic activities, causing large-scale impacts on ecosystems, such as the degradation of soil, pollution of water resources and loss of biodiversity. The magnitude of the environmental impact of these activities depends on the mining technologies used, the extent of the disturbance, chemical, physical and biological factors, mineralogical composition of the soil, vegetation cover, topography, hydrogeological features, and the recovery method used after mining closure. The surface mining causes the following impacts in the areas [2]: it eliminates vegetation; significantly changes the topography; modifies the structure of the soil, subsoil and the geological column drastically and permanently, and disturbs surface and underground hydrological regimes.

In general, the recovery of mined areas consists of relocating the surface layer of soil after removing the coal cover layer and relocating the geological materials to the previous open pit completing the topographic reconstruction stage of the landscape. However, the removal and deposition of the topsoil layer results in degradation of soil quality. Actually, the soils in these areas should be constructed in order to recover the quality of degraded areas after mining activities. Soil types that use mining residues and other types of anthropogenic residues or artifacts are classified as Technosols [3]. The most extensive technosols related to mined areas are mainly associated with modern landscapes created by surface coal mining and are classified as Spolic Technosols, based on the fact that they contain technogenic artifacts in the form of residues from mining activity [4]. In general, constructed technosols could recover most of the basic soil functions and ecosystem services expected for healthy soils [5].

The construction methods for technosols at the Candiota mine, RS, adopted by CRM (**Figure 1**), sometimes use thin topsoil thickness and, therefore, create technosols with a large amount of overburden materials just below these thin topsoil layers, which promotes greater soil density, lower macro-porosity, and high mechanical resistance to penetration, in addition to a low pH value (around 2.4). These soils are constructed with a profile basically composed of two layers: a surface one of topsoil consisting of the A horizon, often mixed with parts of the B and C horizons, and a subsurface composed of a mixture of overburden materials (saprolites and rocks, including a non-mined coal layer from the cover layer above the explored coal seams) [4]. The main steps involved in the construction of technosols at the Candiota mine are the following: (a) removal of the A, B and/or C horizons of the original soil, which are transported by trucks for final coverage of a topographically leveled area; (b) removal of cover rocks using a high-capacity excavator (dragline); (c) extraction of coal banks; (d) in the pit opened by coal extraction, overburden (a mixture of rocks and unused coal) is deposited, which is leveled by bulldozers during the topographic recomposition of the area; (e) completing the topographic reconstruction of the area, a layer of soil (horizon A and/or B) is deposited (topsoil), removed prior to coal mining [7].

Soil compaction is considered the main physical impact on the construction of these technosols. The high compaction is caused by the intense traffic of heavy machinery during the topographic recomposition of the degraded area and has consequences, above all, on the revegetation of these areas [8–10]. The impact of machine traffic at the time of soil reconstruction cause a 23.53% increase in Bd, by immediately after the first pass of the tractor, causing a 20% decrease in total porosity. As a result of the soil compression process, after 12 tractor passes, the percentage of macroaggregates increased from 22.56 to 36.58% [11]. The transit of large machines promotes densification of the layers, reducing the volume of pores and the moisture content of the soil, favoring the connections between mineral particles [12], restricting soil

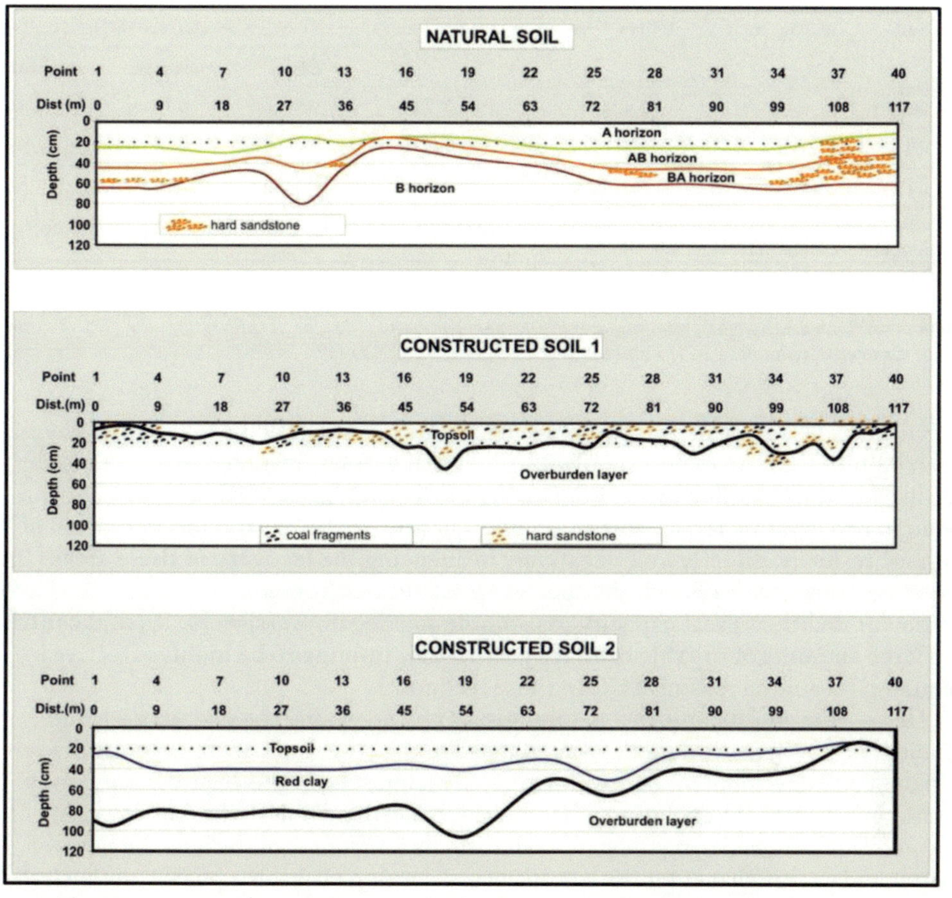

Figure 1.
General sketch of the morphological variation of a natural soil (SN), constructed soil 1 (SC1) and constructed soil 2 (SC2), along profiles and transects, in CRM recovery areas, in Candiota, RS, prior to 1999. Source: [6].

aeration, compromising the root respiration and the development of microorganism populations, due to the significant reduction in soil macro and microporosity. The topographic recomposition and spreading of topsoil with heavy equipment result in the loss of soil structure, determining, among other changes, an increase in soil bulk density (Bd >1.6 Mg m^{-3}) and low rates of infiltration of water into the soil (VIB <5 cm h^{-1}), conditions that favor surface runoff and water erosion [13]. This effect can be observed in **Table 1** where Bd, total porosity (TP) and macro-porosity (Ma) are related to the basic soil water infiltration rates (IR) in technosols (CT1 and CT2) and in natural soil (NS). In relation to natural soil, soils constructed with 1 and 5 years of age have higher density and lower total porosity. The lower IR value in natural soil compared to the technosols is probably due to the Ma index of natural soil, which is relatively lower along practically the entire transect; as its soil density values are lower and its TP values are higher [14]. The very low Ma of the natural soil indicates that this soil must have been degraded by previous use, with excessive cultivation, prolonged periods without vegetation cover and/or trampling by livestock. Furthermore, as it is in a mining front area, this soil may have suffered heavy equipment traffic [6].

Soil	Bd (Mg m^{-3})	PT (m^{-3} m^{-3})	Ma (m^{-3} m^{-3})	IR (mm h^{-1})		
				Mean	Minimum	Maximum
NS	1.42 b	0.4042 a	0.0120 c	0.59 b	0.23	1.10
CT1	1.61 a	0.3510 b	0.0438 b	8.69 a	0.78	35.98
CT2	1.59 a	0.3495 b	0.0577 a	7.33 a	1.20	27.34

Values followed by the same letter in the column do not differ significantly at the α = 0.05 level, using the Tukey test.
Source: Authors, adapted from [6, 14].

Table 1.
Mean of soil bulk density (Bd), total porosity (TP), macro-porosity (Ma), basic infiltration rate (IR) of natural (NS), constructed technosol 1 (CT1) and constructed technosol 2 (CT2) soils.

Over the last 20 years, several physical, chemical and biological variables have been monitored in an experiment with perennial grasses, installed in an area under recovery at the Candiota mine, RS. During this period, great difficulty was encountered in revegetating technosols and, consequently, in the natural incorporation of organic residues, directly and negatively influencing the recovery of these areas [15]. However, during this period, the species *Hemarthria altissima* and *U. brizantha* have demonstrated their great capacity to promote soil decompression [9, 16] and contribute large amounts of organic residues [10, 17–19], proving to be highly effective in reducing Bd and increasing TOC in these technosols.

Despite the enormous effort in monitoring and analyzing the various variables studied, no mathematical model was adjusted to the observed data to describe the behavior and development of the constructed technosols. In this regard, a thermodynamic approach to the evolution of technosols conceptually describes the evolution of these soils [4]. Considering them as open systems far from equilibrium, the flows of matter and energy that enter the system would develop over three phases, as illustrated in **Figure 2**. In the first phase (between t_0 and t_1), they will be at an initial stage (first years) in which the sources of matter and energy (organic matter and microbiota) are in a latent state. Between t_1 and t_2, the flows of matter and energy will be active, in different ways, depending on the type of treatment (T1 – T5), but in a statistically indistinguishable way. From t_2 and t_3, significant statistical differences will be easily evident

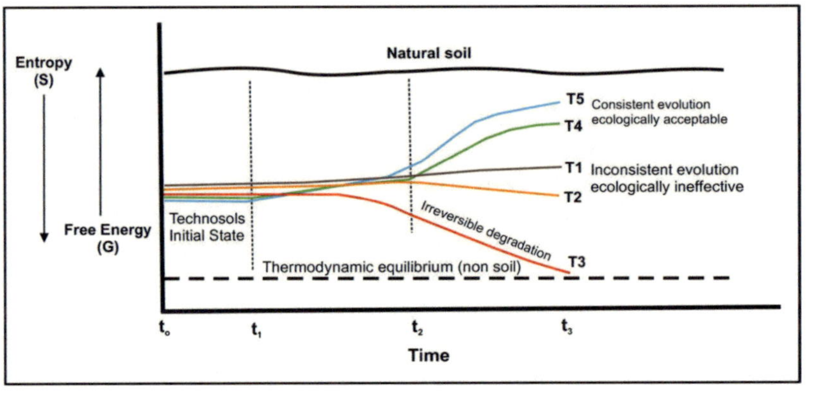

Figure 2.
Hypothetical evolution of technosols constructed in mining areas, under different treatments (T1 to T5), related to different levels of thermodynamic organization. Source: [4].

and technosol will be able to evolve into different future scenarios: (a) irreversible soil degradation (T3); (b) very slow and insufficient evolution to reactivate the local ecosystem (recovery over "time"); (c) the reactivation of flows of matter and energy capable of bringing "life" back to the location (ecosystem recomposed at an acceptable level).

In general, the studies carried out to date in the experimental area of the Candiota mine start from the perspective of a future agro silvopastoral use of the recovered areas. Moreover, technosols constructed in mined areas have a great potential to accumulate significant amounts of carbon, although their ability to function as large-scale carbon sinks and their implications for the global greenhouse gases (GHGs) reduction effort have not yet been assessed. Currently, the enormous losses of carbon stocks, due to changes in land use and management, induced by human activities [20], where mining activities stand out, have led governments to develop sustainable management policies in order to face potential threats of soil degradation, so that their ability to provide essential services is protected or improved [21]. On the other hand, the increase in the concentration of GHG in the atmosphere, especially in recent decades [22], has stimulated the development and consolidation of technologies that allow soils to act as drains for atmospheric C [23].

From this perspective, the recovery of areas degraded by mining activities has received, in recent decades, great attention due to its possible role in combating global warming. Several studies emphasize the interrelationship between the environmental recovery of these areas, C sequestration and nutrient reservoirs [2, 5, 24–28]. Technosols revegetated after mineral exploration can be a significant sink of atmospheric CO_2 through the formation of soil organic matter (SOM) and accumulation of aboveground biomass. In their initial stage of development, technosols offer excellent opportunities for recycling organic waste and maximizing their potential as a carbon sink in relation to agricultural soils, through the adoption of construction and management techniques that enable high rates of carbon sequestration. Proper revegetation of mined areas is an important process to sequester large amounts of atmospheric C and return the land to a stable state [29]. The vegetation of these areas is one of the factors that most contributes to increasing the speed of soil formation in the initial stages of pedogenesis in technosols [30]. Increasing organic matter is an essential factor when seeking to recover soils from areas degraded by mining. The SOM has long been recognized as a universal indicator of soil quality, since it interacts with several soil constituents, influencing water retention, aggregate formation, soil density, pH, cation exchange, nutrient mineralization, sorption of pesticides and other agrochemicals, infiltration, aeration and biological activity of the soil [31].

This study aimed to evaluate the evolution of organic matter indices incorporated by perennial grasses, over 15 years, in a technosol constructed in 2003, as well as its dynamics and interaction with other soil recovery factors and obtain an estimate of carbon stocks incorporated by *H. altissima* and *U. brizantha*.

2. Material and methods

The experiment was installed in a technosol constructed by Companhia Riograndense de Mineração (CRM), in an area under recovery, located in the municipality of Candiota, RS, at geographic coordinates 31° 33′ 55.5″S, 53° 43′ 30, 6″W and altitude of 230 m. The region has a humid subtropical climate, type Cfa [32], with an average annual temperature of 17.2°C, with the highest average temperatures in February (24.2°C) and the lowest in July (12.2°C). The absolute maximum temperature recorded was 45°C and the minimum −2.0°C. The Candiota region has an average annual rainfall of 1599.9 mm, well distributed throughout the year, with a monthly average of 133.3 mm. The highest average precipitation values were observed in the months of April (173.2 mm) and October (176.0 mm). The months of January, February, March, November and December present greater evapotranspiration than precipitation, as the average temperature of these months is higher due to the summer season (**Figure 3**).

The results of the climatological water balance [34] highlighted the months of January, February, March, November and December as having a water deficit while the remaining months had a water surplus (**Figure 4**). The water deficit value in the period was 167.0 mm for the period of a climatological normal (1991–2020), verified between the months of November to March, with values in the following order of magnitude: November (69.7 mm); December (38.6 mm); January (24.4 mm); February (19.4 mm) and March (14.9 mm). The month of December stands out as having the lowest monthly rainfall (114 mm), while the highest average rainfall rates recorded in the period 1991–2020 occurred in the months of April and October, and the excess water calculated in the present study it was 324.3 mm.

The surface mining developed by CRM, in Candiota, RS, impacts an area of 750 ha, of which approximately 610 ha has been topographically recomposed with topsoil (mixture of horizons A, B and C) and 690 ha are under vegetation recovery [35]. The mineral exploration area is divided into 15 grids where, in grid VII, the studied experimental area was installed.

The technosol in the experimental area is characterized by a under layer formed by overburden materials (fragmented mudstones, shales, carbonaceous shales and sandstones removed by detonation of the coal seam cover). The top layer comes from a natural soil from a pre-mined area, classified as Rhodic Lixisol, clayey textural class, dark red color (2.5 YR 3.5/6) and low organic matter content (1.15%) [6]. This top layer (approximately 15–20 cm) is predominantly formed by B horizon mixed with A horizon, removed prior to coal extraction and relocated again as topsoil in areas undergoing recovery [36, 37].

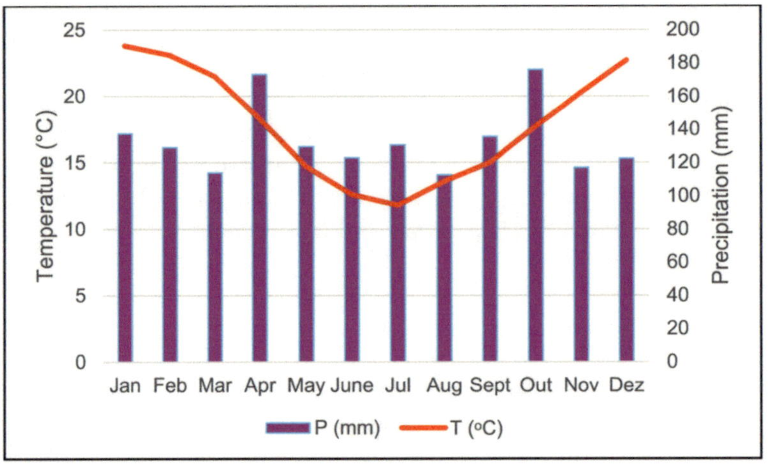

Figure 3.
Annual thermopluviometric variation for the municipality of Bagé, RS, from 1991 to 2020. Source: Authors. Data: [33].

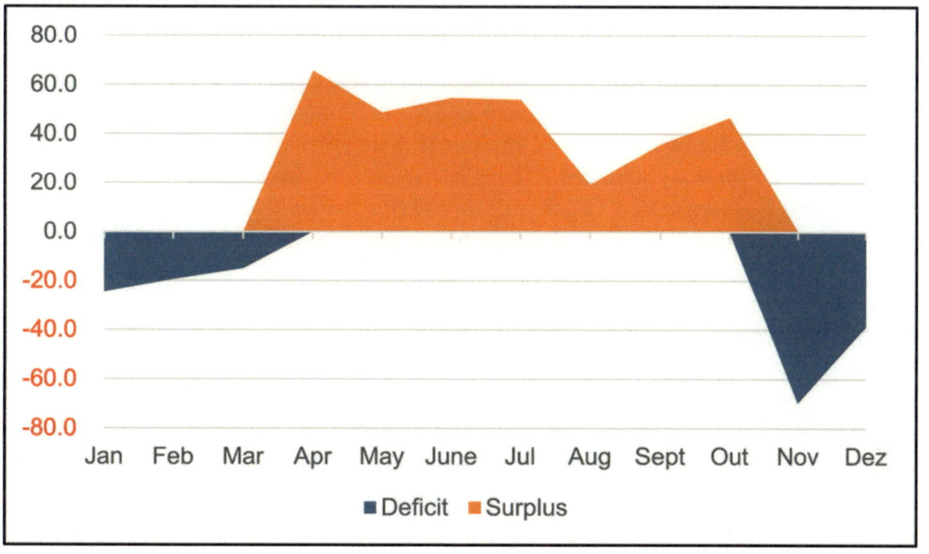

Figure 4.
Water surpluses and deficits (mm) in the municipality of Bagé, from 1991 to 2020. Source: Authors. Data: [33].

Before the experiment was set up, in September 2003, the soil in the area was scarified to a depth of 0.10 to 0.15 m. Next, 10.4 t ha^{-1} of limestone was applied and incorporated, and 900 kg of mineral fertilizer in the 5-20-20 formula, according to the results of the soil analysis. Nitrogen fertilization with ammonium sulfate at a dose of 40 kg ha^{-1} and manual weeding with a hoe were also carried out whenever invasives appeared in the area. Annual fertilizations were carried out on all plots, applying 350 kg ha^{-1}, with formulation 5-30-15 and 250 kg ha^{-1} of ammonium sulfate [37].

Between November and December 2003 different species of cover plants (annual and perennial) were sown. The experiment was installed in plots of 20 m^2 (5 m × 4 m), in a randomized block design, with divided plots, with four replications. Summer crops were arranged in the plots and winter crops in subplots [38]. Perennial summer grasses were used in the plots, with a summer period in winter, single or intercropped with Pinto Peanut (*Arachis pintoi*), composing the following treatments (**Figure 5**).

- T1 – *Hemarthria altissima* (Poiz.) Staff & C. E. Huhh,
- T2 – *Cynodon dactylum* (L.) Pers. cv. Tifton 85 + *Arachis pintoi*,
- T3 – *Hemarthria altissima* (Poiz.) Staff & C. E. Huhh + *Arachis pintoi*,
- T4 – *Paspalum notatum* Flüggé Var. *sausae* Pacodi cv. Pensacola,
- T5 – *Cynodon dactylum* (L.) Pers. cv Tifton 85,
- T6 – *Paspalum notatum* Flüggé Var. *sausae* Pacodi cv. Pensacola+ *Arachis pintoi*,
- T7 –*Urochloa brizantha*
- T8 – Natural soil.

The *A. pintoi*, despite initial development, had its growth reduced over time and practically disappeared in the third year (2006). The same happened with the *Festuca arundinacea* and *Lotus pedunculatus* [38]. With this, the experiment was reduced to only single poaceas: *H. altissima* (T1), *Cynodon dactylum* (T4), *Paspalum notatum* (T5), *U. brizantha* (T7) and the control treatments [37]. The treatments were named as follows: *H. altissima* (T1), *Paspalum notatum* (T2), *Cynodon dactylum* (T3) and *U. brizantha* (T4). For comparative purposes, in areas adjacent to the experiment, the same soil constructed without vegetation cover and samples of natural soil, from the mining front, occupied by native vegetation (shrub and herbaceous) were used.

In the following years (2004, 2005 and 2006), the area was cultivated with black oat (*Avena strigosa*) in the winter and, every summer, treatments were again attempted to be implemented with different species of single and/or intercropped cover crops. In May 2007, when oats were planted, the area was fertilized again with a dose equivalent to 750 kg ha^{-1} (formula 5-20-20). During the entire period, nitrogen was occasionally applied with ammonium sulfate [36].

For 15 years, from May 2004 to April 2018, several works carried out in the experimental area monitored and evaluated the recovery of soils degraded by coal mining in Candiota [7, 37–41]. During this period, several variables were evaluated that describe physical, chemical and biological characteristics of the soil, two of which were considered in this study: soil density (Bd) and total organic carbon content (TOC), both measured in the 0.00–0.10 m layer. Of the species implemented and monitored since 2003, the Poaceae *H. altissima* and *U. brizantha* were selected for this study. The selection of these species prioritized the amount of biomass produced and their adaptation to local soil and climate conditions. This is because the region of Candiota, RS, experiences periods of water deficit in the

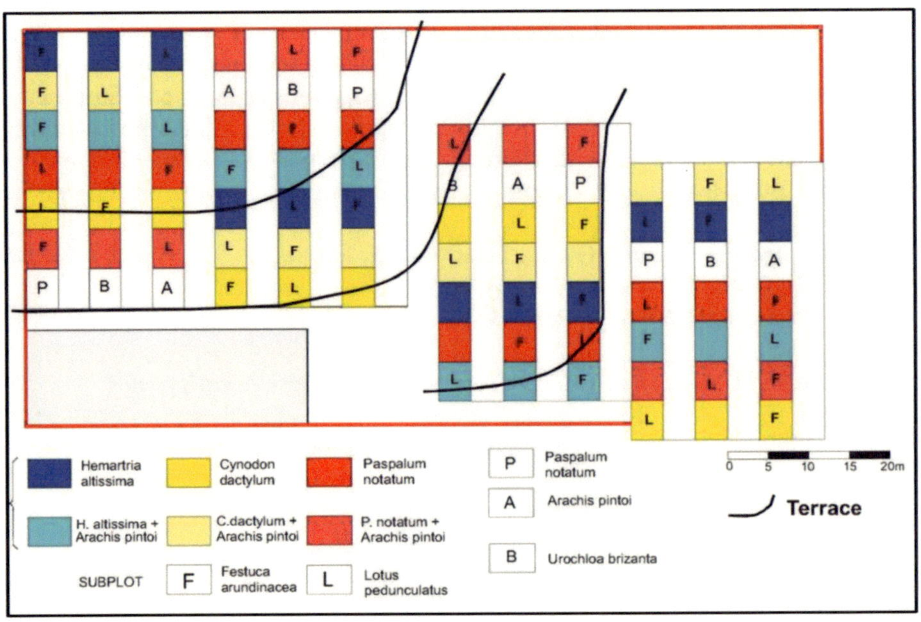

Figure 5.
Sketch of the experimental area, implemented between November and December 2003, in the CRM mining area, in Candiota, RS. Source: Authors.

summer and low temperatures in the winter. Furthermore, they must withstand sites of varying alkalinity and acidity due to the use of limestone and the pyrite oxidation process, which may be mixed with the source material that serves as topsoil.

Fast-growing grasses such as *H. altissima* and *U. brizantha* have a root system that provides better soil coverage rates and can favor the recovery process of areas degraded, preventing soil disintegration [12]. *U. brizantha* is a tropical perennial grass, with loose tufts with short rhizomes and erect or slightly decumbent stems 60–150 (<200) cm high, used as a soil cover to control erosion. Widely distributed, it is best adapted to the humid and sub-humid tropics with average annual precipitation of 1500–3500 mm, but it also grows in more arid tropical regions, with precipitation just below 1000 mm, and can withstand dry seasons of 3 to 6 months. The *U. brizantha* grows in a wide range of well-drained, medium to clayey soils with pH 4–8. It is tolerant to high concentrations of Al^{+3} and is often found in soils with pH < 5.5. Mn tolerance varies between ecotypes, although it can survive in low fertility soils, it requires medium to high soil fertility to be productive. The *H. altissima* is a perennial grass with short rhizomes; culms loosely tufted or prostrate to decumbent, 100–250 cm long, 2–4 mm in diameter, rooting at lower nodes, rising to 30–80 (<160) cm tall. The *H. altissima* grows in soils of any texture, if the humidity is adequate. It produces few seeds, so it is propagated through seedlings (stolon cuttings) planted in moist soil. It tolerates acidic soils up to pH 4.5, but performs best between 5.5 and 6.5 and has a high level of tolerance to excess Al and Mg. It is found in flooded areas, swamps, lakes, however, it can withstand short periods of seasonal drought. It has high growth speed and high biomass addition potential for longer periods than other tropical grasses. As it is found naturally between latitudes of 40°N and 34°S, it tolerates heat (optimal development between 31 and 35°C), frost and cold temperatures of up to −10°C. Its tolerance to low pH and high Al and Mn could be usefully applied to the revegetation of acid mine waste where moisture conditions are suitable [42].

2.1 Statistical analysis

The data considered in this study refer to the Bd and TOC variables, evaluated in plots covered with the grasses *H. altissima* and *U. brizantha*, over a period of 15 years (2003 to 2018). As variable assessments were repeated over time at intervals of varying amplitude, time was measured in months. Thus, six levels were considered for the Time factor: 5, 41, 78, 103, 132 and 175 months after the installation of the experiment.

The statistical analysis of the data comprised the techniques of analysis of variance, to test the significance of the main effects and the interaction of the factors Grass and Time, and polynomial regression analysis to discriminate the variation of the Time factor. For all tests, the significance level $\alpha = 0.05$ was adopted.

Polynomial regression models (Eq. 1) were adjusted for TOC and Bd data in order to identify those that best describe the behavior of these variables over time. The analyses were processed using the statistical analysis system for Windows – Winstat [43].

Polynomial regression is used to describe the relationship between the predictor variable x (Time) and the response variable y (TOC or Bd) as a polynomial of the nth degree in x [44, 45]. The generic form of the nth order polynomial regression equation with a single predictor variable can be represented in the following form:

$$y = \beta_0 + \beta_1.x + \beta_2.x^2 + \ldots + \beta_n.x^n + \varepsilon \quad (1)$$

Where: y is the response variable; x is the predictor variable, β_i are model parameters and ε is the random error.

Estimates of total organic carbon stocks (SCS), stored in soils under *H. altissima* and *U. brizantha* treatments, were calculated [46] using the native soil density of the mining front area as a correction factor.

3. Results and discussion

The results of the analysis of variance, for the TOC and Bd variables, indicate that the main effect of the Time factor was highly significant ($p < 0.0001$), while the main effect of the Grass factor and the interaction between the Grass and Time factors were not significant (**Table 2**). The absence of interaction indicates that the behavior of the response variable over time is the same for both grass species and that the effect of the Grass factor is the same at all levels of the Time factor. The significance of the main effect of the Time factor indicates that the polynomial regression adjustment must be carried out for the marginal means of the Time factor levels.

3.1 The evolution of the technosol over 15 years of monitoring

The polynomial regression adjusted to the observed data is presented in **Figure 6**. Two criteria are used to choose the polynomial regression model: the significance of the model and the magnitude of the value of the coefficient of determination (R^2). For the TOC variable, the linear ($p < 0.0001$) and cubic ($p = 0.0099$) models were highly significant. The adjustment of these models to the observed data obtained indices of $R^2 = 75.24$ (Eq. 2) and $R^2 = 98.89$ (Eq. 3), respectively. The high value of R^2 verified for the cubic model indicates this model as an excellent descriptor of the behavior of TOC data throughout the evolution of the technosol.

$$\hat{\mu} = 2.8461 + 0.09487x \quad (2)$$

$$\hat{\mu} = 4.5866 + 0.1670x - 0.0024x^2 + 1.2059.10^{-5}x^3 \quad (3)$$

The **Figure 6** shows the linear and cubic models, adjusted to data observed between 2004 (5 months) and 2018 (175 months). In 15 years of recovery, the accumulation of organic matter (TOC) in treatments cultivated with *H. altissima* and *U. brizantha* increased by 465%. The initial TOC values after 5 months of implementation of the experiment were 5.36 g kg^{-1} and after 175 months they reached an index of 24.94 g kg^{-1}, higher than that found for natural soil (22.72 g kg^{-1}). It can also be seen, in **Figure 7**, that at 175 months some of the TOC observations exceeded the value of 30 g kg^{-1}.

The development of the polynomial model comprises two subsequent waves, each consisting of three sampling campaigns: the first wave is described by data collected at 5, 41 and 78 months and the second wave corresponds to data collected at 103, 132 and 175 months, after the implementation of the experimental area.

Grass	TOC							Bd						
	Time (months)							Time (months)						
	5	**41**	**78**	**103**	**132**	**175**	**Mean**	**5**	**41**	**78**	**103**	**132**	**175**	**Mean**
U. brizantha	5.48	7.98	10.20	8.40	12.60	24.04	*11.45 a*	1.40	1.57	1.37	1.23	1.27	1.33	*1.36 a*
H. altissima	5.22	8.24	8.48	7.67	13.02	24.17	*11.13 a*	1.47	1.57	1.45	1.40	1.36	1.37	*1.44 a*
Mean	5.35	8.11	9.34	8.03	12.81	24.10	11,29	1.44	1.57	1.41	1.31	1.31	1.35	1.40

Note: Means followed by the same letter in the column do not differ from each other, using the F test, at the level α = 0.05.

Table 2.
Means of the variables total organic carbon (TOC) and soil bulk density (Bd), according to the factors Grass and Time.

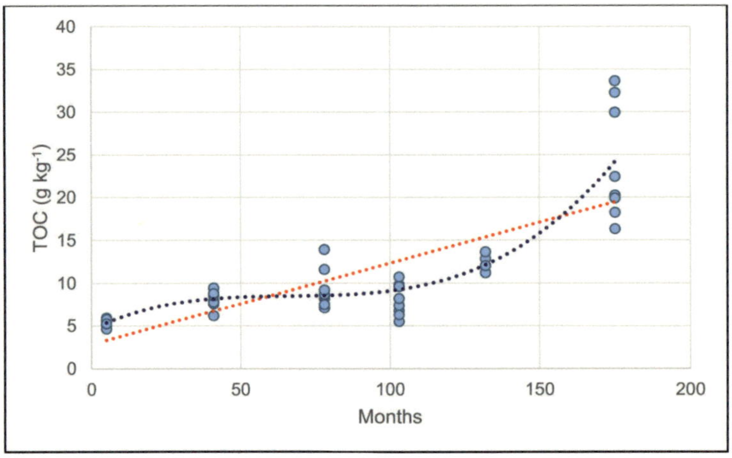

Figure 6.
Linear and cubic models adjusted to TOC data from H. altissima and U. brizantha, over 15 years of observations (175 months), between 2003 and 2018. Source: Authors.

3.2 The first wave of TOC growth: from implementation to 78 months of the experiment

The first wave of TOC growth is comprehended by observations carried out in the first 7 years after the installation of the experiment [38–40]. Considering the thermodynamic model of technosol evolution, this wave would correspond to the first and second phases described [4], where the development of the product takes place from a latent state (1st phase) to a state where statistical differences will be evident (2nd phase). The 1st phase corresponds to the data observed in 2004 [40] and 2006 [38], in the first 48 months of the experiment, when no statistical difference, between the different variables analyzed, was significant. The 2nd phase begins with observations made in 2010 [38], at 78 months and ends at the beginning of the second wave of the model, at 103 months [7]. The estimated TOC values, observed at 5, 41 and 78 months, were 5.22, 8.23 and 8.73 g kg^{-1} respectively. This range of values is considered "low" [47] and is far from the carbon content observed in natural soil taken as a reference (22.72 g kg^{-1}) [6]. However, during this observation period (73 months) there was an increase of 62.8% in the TOC index. However, during this observation period there was an increase of 62.8% in the TOC index. These results correspond to an annual growth rate in the TOC index of 55.4%, in the 1st phase of technosol evolution. This initial effect of the experiment may reflect, in addition to the incorporation of dry matter by cultivated grasses, the numerous attempts to implement the legumes *A. pintoi* and *L. pedunculatus*, as well as other grasses such as *A. strigosa* and *Festuca arundinacea*. It is considered that the various attempts to implement these crops contributed important contributions of dry matter between 2003 and 2006, when attempts to implement legumes as a source of N to the soil were abandoned.

In the 1st and 2nd phases of technosols development, carbon mineralization tends to be very intense right after the incorporation of organic matter into the soil, then progressively slows down until reaching a stage in which little mineralization occurs, and the organic matter that remains is, if not "sequestered", at least stabilized and no longer accessible to microorganisms or their exoenzymes. The kinetics of organic matter in the soil is described by a logarithmic equation where half of the organic

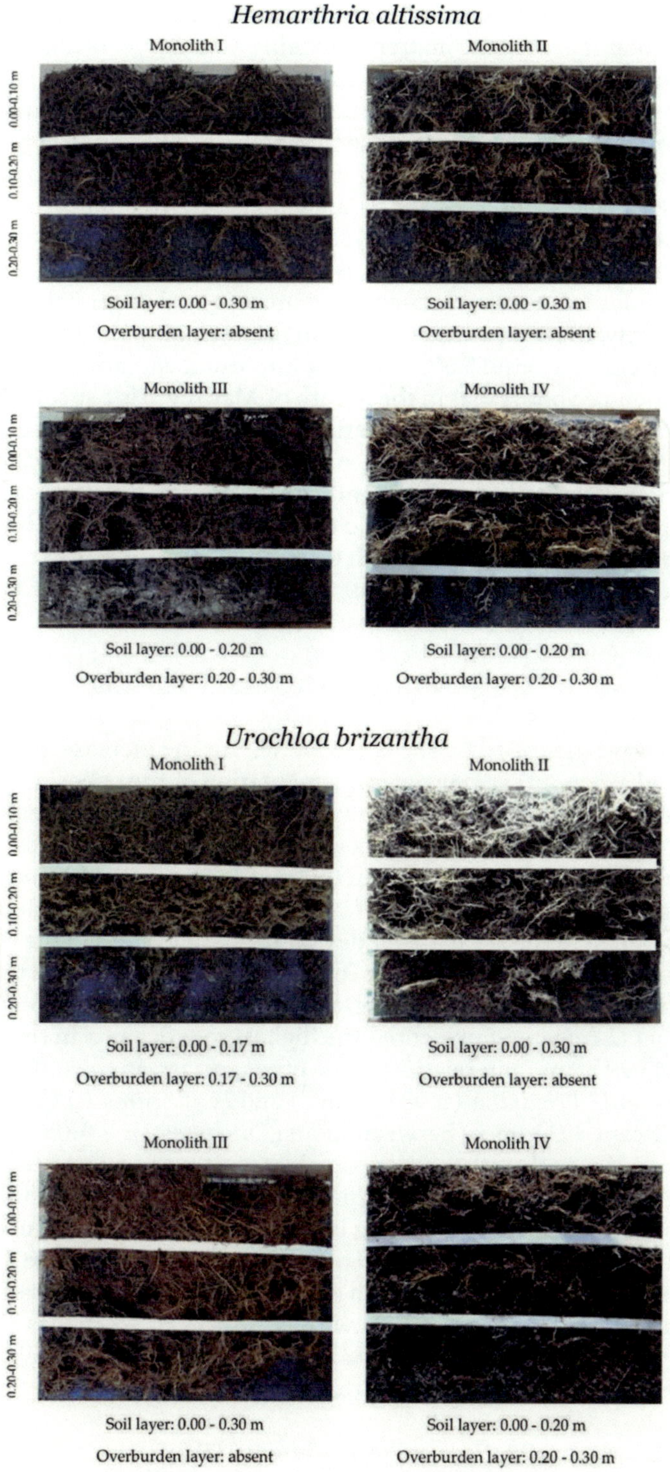

Figure 7.
Qualitative evaluation of the roots at 103 months of revegetation, after washing the soil monoliths removed to a depth of 0.30 m and the thickness of the soil and overburden layer in treatments with H. altissima and U brizantha. Source: [7].

matter added to the soil is mineralized after just over a year, 80% disappears after 7 years and the amount of organic matter remaining after 30 years it is only a tenth of that initially applied [48]. In the case of the incorporation of organic matter into the soil, this behavior can be described by a succession of exponential equations, which can be described by polynomial models such as those shown in the **Figure 6**.

Another important aspect to highlight is the enormous variations observed in several monitored variables, observed in the collections carried out in the months of May 2004, November 2004 and May 2005 [19]. It is believed that such differences are due to climate variations between these months. As can be seen in **Figures 2** and **3**, where regional thermopluviometric variations and climatological water balance are presented, respectively, the temperature variation between May (14.7°C) and November (20.3°C) is, in average, around 5.6°C. Another important climatic aspect associated with the collection periods is that in the month of May there is a surplus of rainfall, while in November the period of water deficit begins in the Candiota region. Thus, under high rates of chemical weathering, such as those found in tropical or subtropical climates, technosols constructed with significant amounts of easily weatherable minerals have the potential to function as efficient carbon sinks [24]. The high potential for soil carbon stocks (SCS) recovery with technosols, particularly in tropical regions, can be attributed to the high input of plant-derived carbon and the strong potential for carbon stabilization through mineral-organic interactions.

3.3 The second wave of TOC growth: from 103 to 175 months

The second wave described by the cubic model, for the increase in TOC, is between 103 and 175 months after the implementation of the experimental area [7, 37, 49]. The average TOC values observed in this period were respectively 9,50, 12,54 e 24,94 g kg^{-1}. During this period, a much more pronounced growth in TOC rates can be seen. The TOC growth rate in this period was 262,4%. This result corresponds to a growth rate of 43,7% $year^{-1}$. This result shows that the deposition of plant residues, added to the greater root development of grasses in the 0.00–0.10 m layer, contributed to the greater increase in TOC on the surface of the technosol. When evaluating the root system of the grasses established in the experiment, it was found that *U. brizantha* presented the highest root density, in the layer of 0.00–0.10 m (13.29 kg m^{-3}) in relation to the other species. In the 0.00–0.10 m layer, the root volume of *U. brizantha* (0.032 m^3 m^{-3}) and *H. altissima* (0.023 m^3 m^{-3}) did not show significant differences between them (Tukey test, α = 0.05), but they were significantly superior to the other grasses evaluated. It is considered that the greater concentration of roots on the surface of the constructed soil may be due to physical or chemical limitations of the underlying layers at 0.10 m, which did not allow for a more uniform development of roots in depth, as can be seen in **Figure** 7 [7].

After 132 months of revegetation with grasses, the density of mites and springtails was significantly higher in the treatment with *H. altissima* [37]. It was observed that lower organic carbon values directly affect the permanence of soil mesofauna organisms in the different treatments in the area under study.

3.4 Effects of increasing TOC on soil bulk density (Bd)

The evolution over the 15 years of monitoring Bd in relation to TOC can be seen in **Figure 8**. The grasses did not show significant differences between them in reducing Bd over this period. The linear regression adjusted to the observed Bd data

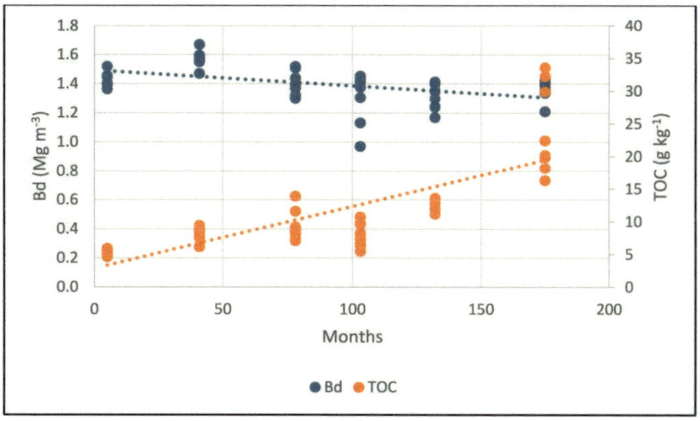

Figure 8.
Variation over 15 years of observation in soil bulk density (Bd) and total organic carbon (TOC) in treatments under cultivation of H. altissima and U. brizantha, in the experimental area of Candiota, RS. Source: authors.

resulted highly significant over time (p < 0,0001), with a coefficient of determination R^2 = 47.57 (Eq. 4). The Bd estimated by this model reduced from 1.49 g kg^{-1} at 5 months (2004) to 1.30 g kg^{-1} at 175 months (2018), representing an 83% reduction in relation to the initial values observed.

$$\hat{\mu} = 1.4945 - 0.0011x \qquad (4)$$

The linear models show that, as the technosol evolves over time, TOC rates increase and Bd decreases. These results agree with other studies [49, 50] that found increases in TOC levels and reductions in Bd as recovery years progressed in different soils, regardless of the type of vegetation. Contrary to some observations [51], the overall relationship between compaction-induced changes in C mineralization and Bd resulted in linear regressions with opposite directions. This means that as the amount of carbon incorporated into the soil increases, the soil density reduces. This is mainly due to the root growth of the grasses under study which, as they increase their density, simultaneously promote the breakdown of aggregates formed by intense soil compaction and increase organic carbon levels through root decomposition. It is noted that, as TOC stocks increase, Bd reduces its intensity. These results converge with other studies that found that vegetation development is strongly correlated with improvements in organic matter and soil bulk density [52, 53].

Although the differences were not significant over the monitoring period, *H. altissima* presented the highest TOC contributions, while its ability to reduce Bd was relatively lower when compared to the reduction promoted by *U. brizantha*. In 2004, the average soil Bd under *H. altissima* was 1.47 Mg m^{-3} and, in 2018, it resulted in 1.45 Mg m^{-3}, while *U. brizantha* presented an average Bd of 1.40 Mg m^{-3} in 2004 and, in 2018, resulted in 1.19 Mg m^{-3}. In 2004, *U. brizantha* showed significant differences in Bd, in the 0.00–0.10 m layer, in relation to other cultivated grasses [39], highlighting the capacity of its root system to occupy the pore spaces of highly compacted soils.

3.5 Technosols in coal mining areas as carbon sinks

After 15 years of recovery with perennial grasses, it was observed that soil carbon stocks (SCS) varied from 5.42 Mg ha^{-1} (5 months) to 22.39 Mg ha^{-1} (175 months) in treatment under *U. brizantha* and from 5.16 Mg ha^{-1} to 22.51 Mg ha^{-1} under *H. altissima*, with no statistical difference between treatments (**Figure 9**). After 175 months of implementation of the experiment, *H. altissima* and *U. brizantha* increased the SCS, on average, by 17.60 Mg ha^{-1} (425%), in relation to 2004. It is noteworthy that throughout the first wave of technosol development (5 to 103 months), the average SCS growth was 0.20 Mg ha^{-1} year^{-1} (16% year^{-1}), a period that corresponds to the 1st and 2nd phases of technosol evolution [4]. Between 103 and 132 months, the SCS increased at a rate of 2.34 Mg ha^{-1} year^{-1} (68% year^{-1}) when, there is a new increase in the growth rate, which reaches 3.08 Mg ha^{-1} year^{-1} at 175 months (55% year^{-1}). These results highlight the great capacity that these grass species have to improve organic matter levels in the soil, both due to their large production of dry matter and their root system.

In Europe and the United States, rates between 0.2 and 2.0 Mg C ha^{-1} year^{-1} were used to predict the effects of reclamation of mined land. Accumulation rates in permanent pastures were relatively higher than other crops, ranging from 0.48 to 1.1 Mg ha^{-1} year^{-1}, in dry tropical lands to cold temperate steppes. Technosols have the potential to sequester carbon and incorporate it in the form of organic matter into the soil, especially in the first years after the start of recovery of the degraded area and may exceed the organic carbon stocks of the original soil [28].

In India, revegetated mined areas compensate 1.20 Mg C ha^{-1} year^{-1} by CO_2 sequestration in the soil, while the total compensation potential through soil, biomass and litter mass is 9.36 Mg ha^{-1} year^{-1}. Therefore, technosols can be considered significant sinks of atmospheric CO_2 through vegetation development, together with ecosystem redevelopment and soil formation. Revegetation of mining areas deserves serious consideration due to its potential to sequester SCS long enough to offset C emissions. Successful revegetation can also serve as a cost-effective way to mitigate the growing problem of CO_2 accumulation in the atmosphere. However, a long-term study on the productivity of these soils is necessary to identify sustainable management options and evaluate C sequestration over a long period of time [27].

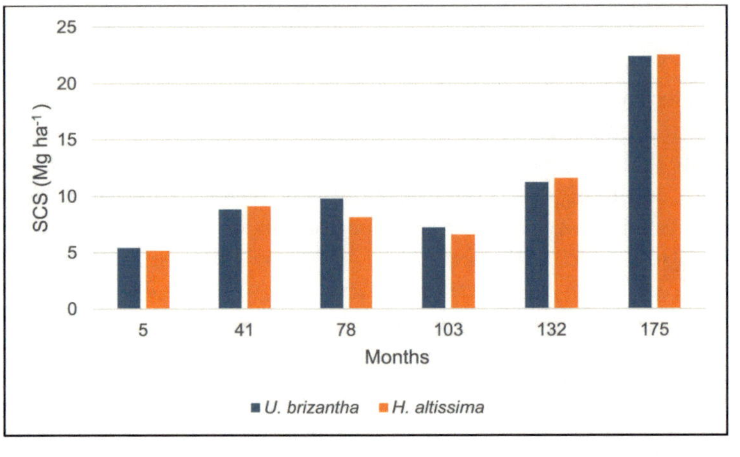

Figure 9.
Evolution of organic carbon stocks (Mg ha^{-1}), in the 0.00–0.10 m layer, produced by H. altissima and U. brizantha, over 15 years (175 months) of recovery of a constructed technosol in surface mining coal area, in Candiota, RS. Source: authors.

4. Final considerations

Several studies have sought to develop quality indices for technosols constructed in mined areas with the aim of investigating the influence of the time factor on the recovery of these areas [54, 55]. In general, these indices are made from chronosequences with different technosols constructed at different times and different source materials and types of waste. In this way, quality indices linked to different variables can be evaluated using different regressive mathematical models. Our study, differing from these, does not require such an approach, as it was implemented in an area intended for research, with experimental control in a randomized block design [16]. The same technosol has been monitored over time and its properties compared to a "natural soil", which makes the data obtained more reliable and the conclusions more robust and scientifically significant. In this study, in addition to generating a model, with high statistical significance and excellent adjustment to observed data, capable of describing the evolution of a technosol, over 15 years of observations, the potential of these technosols to contribute to the global effort to reduce of GHG.

The large impacts generated by coal mining for thermal power generation in terms of GHG emissions can be partially mitigated with technosols properly designed for this purpose and with vegetation management techniques that promote the rapid incorporation of waste into the soil [55]. In addition to topographic recomposition and revegetation of the degraded landscape, with a view to reducing damage caused by erosion processes to the new environments generated, technosols have great potential for use as carbon sinks.

Past precedence and research results show that the reclaimed soil properties in coal mining areas even after several years of reclamation are still evolving [16]. Hence, the most economical/cost effective and practical approach to bring the degraded soil to near natural soil is to '*leave the land as it is*' (i.e., *in situ* and undisturbed). Slowly, over a period of time the cultivability of soil is restored automatically through many naturally active agents, carriers and natural unclaimed processes dominant in the mining areas. However, it is important to highlight that, in the beginning, correcting soil acidity and fertilization are essential for the implementation of crops to be used in the recovery of degraded areas.

Although the regression analysis pointed to the cubic model as an excellent descriptor of the evolution of the technosol in its first 15 years, other models will probably be developed as the years of monitoring progress. Considering the behavior presented by the set of observed data and the trend of accelerated growth in TOC indices, it is believed that the best descriptor of the long-term evolution behavior of technosols is best described by a succession of LRP-type regression models (linear response plateau). The LRP model is a type of regression segmented into two parts, where in the first part the response is described by a simple linear regression model and the other is described by one parallel to x (plateau), that is, where the response is constant [56]. In our chronosequence, this model would adapt to the set of data obtained in the first four evaluations [7, 38–40], of which the first two would correspond to simple linear regression and the remaining to Plateau.

5. Conclusions

The cubic model adjusted to data observed over 15 years of technosol monitoring, obtained high statistical significance and an excellent coefficient of determination, being able to robustly express the evolution of the technosol.

The grass species evaluated (*H. altissima* and *U. brizantha*) did not show significant differences between them, over 15 years of observations, in the incorporation of TOC, in soil carbon stocks and in the reduction of soil density.

Throughout this period, the TOC content in the surface layer of the soil exceeded the levels observed in the natural soil used as a comparison parameter.

The results presented in this study corroborate the importance of using perennial grasses as a technology capable of leading the evolution of technosols constructed in mining areas to the higher stages of the environmental recovery process.

Acknowledgements

The authors would like to acknowledge Companhia Riograndense de Mineração (CRM), Rede Carvão (Brazilian Coal Network), IBAMA (Brazilian Institute of the Environment and Renewable Natural Resources), CAPES and CNPq for logistical and financial support.

Conflict of interest

The manuscript is original, has not been published before, and is not being considered for publication elsewhere in its final form neither in printed nor in electronic format and does not present any kind of conflict of interests. The publication has been approved by all coauthors as well as by the responsible authorities at the institute where the work has been carried out.

Author details

Luís Eduardo Torma Burgueño[1*], Luiz Fernando Spinelli Pinto[2], Lizete Stumpf[2], Clause Fátima de Brum Piana[3], Maurício Silva de Oliveira[1], Maurizio Silveira Quadro[4] and Pablo Miguel[2]

1 Soil and Water Management and Conservation Graduate Program, Federal University of Pelotas, Brazil

2 Agronomy Course, Soil Departament, Federal University of Pelotas, Brazil

3 Technological Development Center, Federal University of Pelotas, Brazil

4 Engineering Center, Federal University of Pelotas, Brazil

*Address all correspondence to: eduardo.burgueno@gmail.com

References

[1] Brasil, 2023. Ministério de minas e energia. Empresa de Pesquisa Energética – EPE. Balanço Energético Nacional – BEN. Relatório Síntese 2023; Ano Base 2022. 65 p. Available from: https://www.epe.gov.br/sites-pt/publicacoes-dados-abertos/publicacoes/PublicacoesArquivos/publicacao-748/topico-687/BEN2023.pdf [Accessed date: February 19, 2024]

[2] Shrestha RK, Lal R. Ecosystem carbon budgeting and soil carbon sequestration in reclaimed mine soil. Environment International. 2006;**32**:781-796. DOI: 10.1016/j.envint.2006.05.001 [Accessed date: April 11, 2024]

[3] WRB. Base de Referência Mundial para Recursos de Solo, 2014. Sistema Internacional de Classificação de Solos para Nomeação de Solos e Criação de Legendas para Mapas de Solos. Roma: FAO; 2014. 192 p. Relatórios Mundiais de Recursos do Solo. p. 106 [Accessed date: June 15, 2023]

[4] Pinto LFS, Fernandes FF, Pauletto EA, Stumpf L. Recuperação de solos degradados pela mineração: da teoria termodinâmica à prática nas áreas regeneradas da mina de carvão de Candiota, RS. In: Alba JMF, editors. Recuperação de áreas mineradas. 3ª ed. Brasília, DF: Embrapa; 2018. 456 p. rev. e ampl

[5] Ruiz F, Cherubin MR, Ferreira TO. Soil quality assessment of constructed Technosols: Towards the validation of a promising strategy for land reclamation, waste management and the recovery of soil functions. Journal of Environmental Management. 2020;**276**:111344. [Accessed date: October 12, 2023]

[6] Nunes MCD. Condições físicas de solos construídos na área de mineração de carvão de Candiota-RS (Dissertação Mestrado em Ciências). Brazil: Programa de Pós-graduação em Agronomia, Universidade Federal de Pelotas, Faculdade de Agronomia Eliseu Maciel, Depto. de Solos; 2002, 131 p. Pelotas, RS, 2006, 120 p [Accesssed date: June 14, 2023]

[7] Stumpf L. Desenvolvimento radicular de gramíneas perenes e sua eficiência na recuperação de atributos físicos de um solo construído compactado em área de mineração de carvão Tese (Doutorado). Pelotas: Programa de Pós-Graduação em Agronomia, Faculdade de Agronomia Eliseu Maciel, Universidade Federal de Pelotas; 2015. 122 p [Accesssed date: February 18, 2024]

[8] Miguel P, Stumpf L, Pinto LFS, Pauletto EA, Rodrigues MF, Barboza LS, et al. Physical restoration of a minesoil after 10.6 years of revegetation. Soil and Tillage Research. 2023;**227**:105599. DOI: 10.1016/j.still.2022.105599 [Accesssed date: September 13, 2023]

[9] Pauletto EA, Stumpf L, Pinto LFS, Silva TS, Ambus JV, Garcia GF, et al. Reclamation of a degraded coal-mining area with perennial cover crops. Revista Brasileira de Ciência do Solo. 2016;**40**:e0150482. DOI: 10.1590/18069657rbcs20150482 [Accesssed date: August 30, 2022]

[10] Stumpf L, Pauletto EA, Fernandes FF, Suzuki LEAS, Silva TS, Pinto LFS, et al. Perennial grasses for recovery of the aggregation capacity of a reconstructed soil in a coal mining area in Southern Brazil. Revista Brasileira de Ciência do Solo. 2014;**38**:327-335 [Accesssed date: August 23, 2022]

[11] Barboza FD, Stumpf L, Pauletto EA, De Lima CLR, Pinto LFS, Jardim TM,

et al. Impact of machine traffic events on the physical quality of a minesoil after topographic reconstruction. Soil and Tillage Research. 2021;**210**:104981 [Accesssed date: July 19, 2023]

[12] Reis DA, Lima CLR, Pauletto EA. Resistência tênsil de agregados e compressibilidade de um solo construído com plantas de cobertura em área de mineração de carvão em Candiota. Revista Brasileira de Ciência do Solo. 2014;**38**:669-678 [Accesssed date: February 24, 2024]

[13] Bitencourt DGB, Pinto LFS, Pauletto EA, Silva MT, Garcia GF. Geração de drenagem ácida e de contaminação por metais pesados em perfis de solos construídos em área de mineração de carvão. Revista Brasileira de Ciência do Solo. 2015;**39**:1-14 [Accesssed date: February 14, 2024]

[14] Guadagnin CA, Pauletto EA, Pinto LFS. Taxa de infiltração estável da água ao longo de transeções em solos construídos na área de mineração de carvão de Candiota – RS. Revista Brasileira de Agrociência, Pelotas. 2007;**13**(3):337-342 [Accesssed date: February 14, 2024]

[15] Guadagnin CA. Infiltração de água ao longo de uma transeção em solos construídos na área de mineração de carvão de Candiota – RS Dissertação (Mestrado em Ciências). Pelotas, 62 p: Programa de Pós-Graduação em Agronomia, área de concentração: Solos. Faculdade de Agronomia Eliseu Maciel, Universidade Federal de Pelotas; 2002 [Accesssed date: April 02, 2024]

[16] Pinto LFS, Stumpf L, Miguel P, Dutra LA Jr, Leidemer JD, Barbosa LS, et al. Reclamation of soils degraded by surface coal mining. In: Soni A, editor. Mining Techniques - Past, Present and Future. IntechOpen; 2021. DOI: 10.5772/intechopen.79241 [Accesssed date: August 31, 2022]

[17] Stumpf L, Pauletto EA, Pinto LFS, Geissler LO, Castilhos DD, Souza DL, et al. Biological and physical quality of a mined soil under revegetation with perennial grasses. Revista Brasileira de Ciências Agrárias. 2018;**13**(1):e5498. DOI: 10.5039/agraria.v13i1a5498 [Accesssed date: August 30, 2022]

[18] Jardim TM, Geissler L, Pauletto EA, Morselli TBGA, Castilhos DD, Pinto LFS, et al. Fauna edáfica de um solo construído em área de mineração de carvão revegetado com diferentes espécies de Poáceas. In: Anais do... V Cbcm – congresso brasileiro de carvão mineral. Associação Brasileira de Carvão Mineral, Criciúma, SC, Brasil. De 29 de maio a 01 de junho. 2017 [Accesssed date: March 02, 2024]

[19] Stumpf L, Pauletto EA, Pinto LFS, Ambus JV, Garcia GF, Silva TS. Características químicas de um solo construído e seus efeitos no desenvolvimento radicular de gramíneas perenes. Revista Brasileira de Ciências Agrárias. 2016;**11**(4):343-349 [Accesssed date: February 14, 2024]

[20] Sanderman J, Hengl T, Fiske GJ. Soil carbon debt of 12,000 years of human land use. Proceedings of the National Academy of Sciences. 2017;**114**:9575-9580. DOI: 10.1073/pnas.1706103114 [Accesssed date: March 17, 2024]

[21] Gregory AS, Ritz K, McGrath SP, Quinton JN, Goulding KWT, Jones RJA, et al. A review of the impacts of degradation threats on soil properties in the UK. Soil Use and Management. 2015;**31**(Suppl. 1):1-15. DOI: 10.1111/sum.12212 [Accesssed date: October 14, 2023]

[22] Marques L. O decênio decisivo: propostas para uma política de sobrevivência. São Paulo: Elefante; 2023, 624 p

[23] Dick DP, Hanke D. A matéria orgânica do solo: propriedades, reações e processos. In: Tiecher T, Silva LS, Martins AP, Mallmann FJK, editors. Química do Solo. Santa Maria: Sociedade Brasileira de Ciência do Solo – Núcleo Regional Sul; 2023. pp. 83-138, 320 p

[24] Ruiz F, Safanelli JL, Perlatti F, Cherubin MR, Demattê JAM, Cerri CEP, et al. Constructing soils for climate-smart mining. Communications Earth and Environment. 2023;**4**:219. DOI: 10.1038/s43247-023-00862-x [Accesssed date: October 12, 2023]

[25] Ahirwal J, Kumar A, Pietrzykowski M, Maiti SK. Reclamation of coal mine spoil and its effect on technosol quality and carbon sequestration: A case study from India. Environmental Science and Pollution Research. 2018;**25**:27992-28003 [Accesssed date: October 12, 2023]

[26] Smith P. Soil carbon sequestration and biochar as negative emission technologies. Global Change Biology. 2016;**22**:1315-1324. DOI: 10.1111/gcb.13178 [Accesssed date: October 24, 2023]

[27] Tripathi N, Singh RS, Nathanail CP. Mine spoil acts as a sink of carbon dioxide in Indian dry tropical environment. Science of the Total Environment. 2014;**468-469**:1162-1171. DOI: 10.1016/j.scitotenv.2013.09.024 [Accesssed date: April 11, 2024]

[28] Ussiri DAN, Lal R. Carbon sequestration in reclaimed minesoils. Critical Reviews in Plant Sciences. 2005;**24**(3):151-165. DOI: 10.1080/07352680591002147 [Accesssed date: October 11, 2023]

[29] Shrestha RK, Lal R. Carbon and nitrogen pools in reclaimed land under forestand pasture ecosystems in Ohio, USA. Geoderma. 2010;**157**:196-205. DOI: 10.1016/j.geoderma.2010.04.013 [Accesssed date: April 26, 2024]

[30] Ferreira TO, Ruiz F, Queiroz HM, Otero XL, Cherubin MR, Lopes TA, et al. Tecnossolos: da pedogênes às aplicações ambientais. In: Nogueira TAR, Cherubin MR, Pereira APA, Tiecher T, editors. Tópicos em Ciência do Solo. Vol. 12. Viçosa, MG: Sociedade Brasileira de Ciência do Solo; 2023. pp. 290-316, 466 p

[31] Cerri CEP, Abbruzzini CBB, Durigan MR, Freitas RCA, Lisboa IP, Oliveira DMS, et al. Estoques de carbono e fluxo de gases de efeito estufa em agrossistemas no Brasil. In: Nascimento CWA, Souza VS Jr, Freire MBGS, Souza ER, editors. Tópicos em Ciência do Solo. Vol. 9. Viçosa, MG: Sociedade Brasileira de Ciência do Solo; 2015. pp. 1-47, 290 p

[32] Köppen W. Climatologia: con un estudio de los climas de la tierra. México: Fondo de Cultura Econômica; 1948. 479 p

[33] Inmet – Instituto Nacional De Meteorologia. Normais climatológicas do Brasil – Período de 1961-2020. Estação Fluviométrica n. 83.980, Bagé, RS. Disponível em: https://portal.inmet.gov.br/normais [Accesssed date: October 11, 2023]

[34] Thornthwaite CW, Mather JR. The Water Balance. Vol. 8(1). Centerton, NJ: Drexel Institute of Technology - Laboratory of Climatology. Publications in Climatology; 1955. 104 p

[35] Companhia Riograndense De Mineração - CRM. Balanço patrimonial do exercício. 2021. Available from: https://crm.rs.gov.br/upload/arquivos/202205/02104808-balanco-2021.pdf [Accesssed date: February 15, 2023]

[36] Holanda-Neto MR. Atributos microbiológicos de um solo construído vegetado com gramíneas após mineração

de carvão em Candiota/RS (Dissertação Mestrado em Ciências). Brazil: Programa de Pós-Graduação em Agronomia. Faculdade de Agronomia Eliseu Maciel, Universidade Federal de Pelotas; 2011, 96 p [Accesssed date: April 02, 2024]

[37] Geissler LO. Atributos físico-químicos e da fauna edáfica como indicadores de recuperação de áreas degradadas pela mineração de carvão (Dissertação Mestrado em Ciências). Pelotas: Programa de Pós-graduação em Manejo e Conservação do Solo e da Água, Universidade Federal de Pelotas, Faculdade de Agronomia Eliseu Maciel, Depto de Solos; 2016, 94 p [Accesssed date: February 10, 2024]

[38] Castro RC. Avaliação temporal de atributos físicos de um solo construído em área de mineração de carvão recuperado com gramíneas perenes (Dissertação Mestrado em Ciências). Pelotas, RS: Programa de Pós-graduação em Agronomia, Universidade Federal de Pelotas, Faculdade de Agronomia Eliseu Maciel, Depto de Solos; 2012. 99 p [Accesssed date: February 24, 2024]

[39] Franco AMP. Caracterização física de um solo construídos na área de mineração de carvão de Candiota, RS. (Dissertação Mestrado em Ciências). Pelotas, RS: Programa de Pós-graduação em Agronomia, Universidade Federal de Pelotas, Faculdade de Agronomia Eliseu Maciel, Depto de Solos; 2006, 120 p [Accesssed date: February 15, 2024]

[40] Gonçalves FC. Efeito de plantas de cobertura sobre os atributos físicos de um solo construído na área de mineração de carvão de Candiota, RS após três anos. Pelotas, RS (Dissertação Mestrado em Ciências). Pelotas: Programa de Pós-graduação em Manejo e Conservação do Solo e da Água, Faculdade de Agronomia Eliseu Maciel, Universidade Federal de Pelotas; 2008, 91 p [Accesssed date: February 14, 2024]

[41] Rodrigues MF. Avaliação da recuperação da estrutura de um solo construído em áreas de mineração de carvão utilizando um método visual (Dissertação Mestrado em Ciências). Pelotas, RS: Programa de Pós-Graduação em Manejo e Conservação do Solo e da Água - Universidade Federal de Pelotas; 2010. 66 p [Accesssed date: April 12, 2024]

[42] Cook BG, Pengelly BC, Schultze-Kraft R, Taylor M, Burkart S, Cardoso Arango JA, et al. Tropical Forages: An Interactive Selection Tool. 2nd and Revised ed. Nairobi, Kenya: International Center for Tropical Agriculture (CIAT), Cali, Colombia and International Livestock Research Institute (ILRI); 2020. Available from: www.tropicalforages.info [Accesssed date: February 23, 2023]

[43] Machado AA. Sistema de análise estatística para Windows (WINSTAT). Pelotas, RS: Universidade Federal de Pelotas, CD; 2001

[44] Montgomery DC, Runger GC. Estatística aplicada e probabilidade para engenheiros. 6ª ed. Rio de Janeiro: LTC; 2018. 629 p

[45] Devore JL. Probabilidade e estatística para engenharia e ciências. São Paulo: Cengage Learning; 2016. 693 p

[46] Fernandes FA, Fernandes AHBM. Cálculo dos estoques de carbono do solo sob diferentes condições de manejo. Vol. 69. Corumbá: Embrapa Pantanal. (Embrapa Pantanal. Comunicado Técnico); 2009. 4 p. Disponível em: http://www.cpap.embrapa.br/publicacoes/download.php?arq_pdf=COT69> [Accesssed date: April 10, 2024]

[47] Sociedade Brasileira De Ciência Do Solo. Manual de adubação e de calagem para os Estados do Rio Grande do Sul e de Santa Catarina. 10ª ed. Porto Alegre:

Sociedade Brasileira de Ciência do Solo. Comissão de Química e Fertilidade do Solo, 2004, 400 p

[48] Berthelin J, Laba M, Lemaire G, Powlson D, Tessier D, Wander M, et al. Soil carbon sequestration for climate change mitigation: Mineralization kinetics of organic inputs as an overlooked limitation. European Journal of Soil Science. 2022, 9 p;**73**:e13221. DOI: 10.1111/ejss.13221 [Accesssed date: November 25, 2023]

[49] Li J, Xin Z, Yan J, Li H, Chen J, Ding G. Physicochemical and microbiological assessment of soil quality on a chronosequence of a mine reclamation site. European Journal of Soil Science. 2018;**69**:1056-1067. DOI: 10.1111/ejss.12714 [Accesssed date: April 10, 2024]

[50] Li X, Lei S, Liu F, Wang W. Analysis of plant and soil restoration process and degree of refuse dumps in open-pit coal mining areas. International Journal of Environmental Research and Public Health. 2020;**17**:1975. DOI: 10.3390/ijerph17061975 [Accesssed date: November 23, 2023]

[51] Beylich A, Oberholzer HR, Stefan Schrader S, Heinrich Höper H, Wilke BM. Evaluation of soil compaction effects on soil biota and soil biological processes in soils. Soil and Tillage Research. 2010;**109**:133-143. DOI: 10.1016/j.still.2010.05.010 [Accesssed date: October 14, 2023]

[52] Feng Y, Wang JM, Bai ZK, Reading L. Effects of surface coal mining and land reclamation on soil properties: A review. Earth Science Reviews. 2019;**191**:12-25. DOI: 10.1016/j.earscirev.2019.02.015 [Accesssed date: July 19, 2023]

[53] Wang S, Cao Y, Pietrzykowski M, Zhou W, Zhao Z, Bai Z. Spatial distribution of soil bulk density and its relationship with slope and vegetation allocation model in rehabilitation of dumping site in loess open-pit mine area. Environmental Monitoring and Assessment. 2020;**192**:1-18 [Accesssed date: October 11, 2023]

[54] Mukhopadhyay S, Maiti SK, Masto RE. Development of mine soil quality index (MSQI) for evaluation ofreclamation success: A chronosequence study. Ecological Engineering. 2014;**71**:10-20. DOI: 10.1016/j.ecoleng.2014.07.001 [Accesssed date: April 11, 2024]

[55] Fernandez MBG, Stumpf L, Miguel P, Pinto LFS, Carlos FS, Schubert RN, et al. Biological quality and organic matter dynamics in minesoil at 18 years reclamation with perennial grasses. Ecological Engineering. 2023;**187**:106866. DOI: 10.1016/j.ecoleng.2022.106866 [Accesssed date: July 19, 2023]

[56] Silva RA. Estatística decodificada. São Paulo: Blucher; 2023. 418 p

Chapter 4

Stability of Crown Pillar during Transition from Open Pit to Underground Mining

Chandrani Prasad Verma and Abhay K. Soni

Abstract

Mineral deposits at shallow depth are on the verge of exhaustion and transiting to underground mining operations, for extraction of deposits at depth has now become inevitable. The depth of open pit mine is generally governed by the stripping ratio and lateral extent of lease. When open pit reaches its ultimate designed depth and the ore is still existing at greater depth transition is must and barrier pillar needs to be left in between both the workings. If underground mining is below an opencast mine then stability of barrier/crown pillar between open pit and underground mining becomes critical especially when non-caving method is to be followed. It has its own implications and maintaining safety of pit and slope along with underground workings becomes a real challenge. Ground monitoring in terms of induced stresses and displacement holds paramount importance. The most popular method in metal mining is Long Hole/Blast Hole Stoping Method in top down sequence. Earlier hydraulic fill was common and is now being slowly replaced by Paste fill. In this chapter a few challenges existing in Indian mining scenario is discussed along with review of methodologies for optimum design of crown pillar. Parameter influence is also explored for its stability.

Keywords: numerical modeling, rock mechanics, ground control, transition, challenges

1. Introduction

The ore that exists at shallow depth is usually extracted by surface mining. However, deposits that lies at greater depth is exploited by underground mining. Most of the opencast mine has to undergo transition from open pit to underground at certain depth when open pit mining is no longer feasible either due to economy or due to limitation in lateral extent. This transition requires a barrier pillar in between open pit and underground mine to isolate one working from another. This barrier is commonly termed as Crown pillar. It is nothing but the ore bearing rock that is left at the bottom of the open pit for support of underground workings and to extract the ore deposit existing at greater depth safely.

This barrier pillar plays a vital role during the entire mining operation and its stability analysis is of utmost important for the protection of both surface mine slopes and underground mine infrastructure. The main purpose of transition crown pillar is to protect the ground surface and underground mine from subsidence and any material inflows like water, soil and rock into the mine while underground mining activity is going on. In mine closure planning too the evaluation of the stability of crown pillar is one of the most important criteria [1]. Many metal mines in India and across the world are at greater depth more than 600 m [2]. Hence, optimization of crown thickness and stability analysis of crown pillar is quite critical for safe underground operations. Preferably, an optimum size of pillar is required to prevent undue loss of ore.

2. Existing methodologies

Apart from a few structural mechanics approach that does not consider rock mass characteristics only two Empirical approaches and Numerical Modeling approach exists for design of Crown Pillar. Initially, the thumb rule was to compute ratio of crown thickness (T) to span (S) for design. A ratio of 1:1 is considered as good rock while thickness to span ratio of 3:1 or more is considered as poor. It is also quoted as Thick cross ratio method by Sun et al. [3]. Later it was modified and linked to rock mass as $T/S = 1.55\ Q^{-0.62}$ where Q is NGI's rock mass quality index. Few structural mechanics based analytical approaches like Slab Beam Theory, Ruppenpit formula, Compressive Arch Theory, Engineering calculation method, etc. have been developed [4] but not that useful for mining scenario as it does not take into account characteristics of rock mass. Software named CPillar is developed by Rocscience that computes FoS of crown pillar using deterministic analysis, however, widely used method of pillar design is scaled span concept developed by Carter et al. [5, 6].

Carter proposed Scaled Span approach long back in late 1980's for surface crown of near surface mined openings. Originally, it included 200 case record of near-surface openings with 30 failure cases. Later in 2002 and 2008 further modifications were made and 114 case studies were added by extending the original data base. New case studies were primarily from shallow dipping situations, including twenty-one additional failure cases for pillar reliability in long term closure planning. These cases included both hard rock situations and soft rock examples including coal [7]. Failure probability concept is also introduced in 2008. In 2014 guidelines for use of scaled span concept was published and experience from 500 cases with over 70 failures were utilized to guide design of crown pillar. Scaled crown span method have been used in a number of cases but inadvertent failure still continues to occur [7]. It probably does not involve crown pillar design during transition phase of opencast mining to underground mining, stress reorientation in rock mass on account of open pit mining and its subsequent effect on crown pillar. It is limited to deposits being mined at very shallow depths by underground mining. Influence of higher working depth, slope stability scenario and simultaneous open pit and underground mining stage where blasting in open pit is likely to disturb integrity of crown pillars is perhaps also not included.

Bakhtavar et al. [8] has developed an empirical relation to estimate crown pillar thickness between open pit and underground mining. The equation has been derived from data sets of a few case studies through multiple regression. But its application in real field has not been found reported. Use of numerical modeling like 3DEC and FLAC3D are also cited in literature. Numerical modeling is sensitive to input parameters and needs to be calibrated. Numerical code RFPA2D is used to analyze the stability of boundary/crown pillar [9]. Failure of rock mass that occurred at Yanqianshan Iron Mine is studied [10] using numerical modeling where the mine was worked by underground mining under an open pit. Kalenchuk [11] analyzed crown pillar behavior through FLAC3D to ascertain crown thickness in a coal mine [12]. Crown thickness for the transition zone of a gold mine was designed using FLAC3D and same model was used for replacement of the crown with artificial pillar [4]. FLAC3D is used to perform parametric study to investigate the effect of fault properties and horizontal-to-vertical in-situ stress ratio on surface crown pillar stability [13].

Based on the analysis of failure cases in most of the surface crown pillars, five pattern of failures mechanisms were also identified by Golder [5, 14, 15] as plug failure, chimneying, caving, unraveling and delamination. Each of this failure represents unique feature and are limited to near-surface mine openings. Failure of the surface crown is a large-scale loss that may result in subsidence or sinkhole formation.

3. Lessons from abroad

A few cases where some kind of crown pillar failure occurred is discussed to visualize the likely issues a mine can face when crown becomes unstable and also the reasons associated with it. Failure that occurred on the north pit wall of Palabora Mine, South Africa includes the collapse of the crown pillar that resulted when block caving was initiated 400 m below the open pit having a depth of 800 m from surface. From modeling study it was predicted that failure is most likely to occur on the north wall, which is mainly controlled by the underground working caving operations and the rest east, south and west walls are appeared to be in stable condition for long time [16]. It is also found that pervasive joint sets in the north wall form wedges that daylight into the caved region below the pit.

Two rock mass collapse was observed in Shirengou iron ore Mine of China, one on account of fault movement and another due to water seepage. The method of working was Short-hole shrinkage stoping method. Although design thickness of crown pillar was around 25–30 m but the actual thickness was quite less mainly due to over excavation and illegal extraction. As a result, a few goaves were just 4–8 m below pit bottom. From the detailed analysis, it was found that fault lying in the crown pillar had undergone movement due to the underground mining activity and it propagated that finally leads to collapse of the crown pillar. Prior to the failure, several abnormalities were observed in the micro-seismic parameters such as abrupt change in energy release and stress levels [17].

In Chuquicamata copper ore mine of Chile method of working used was Panel caving. The main geological aspects was the presence of regional fault and shear zone. The rock mass consists of several discontinuities and the failure was initiated due to the earthquake in December 1967. Due to this earthquake the discontinuities began to open at the south end of east wall. Additionally, work of excavation and the blasting in the mine leads to collapse of the crown pillar in the upper part i.e., north block of the slope. It was found that crown pillar failure occurs in a sequence of events and finally lead to collapse of pillar [18].

Slope wall collapse, water percolation in underground, formation of pothole and movement along fault plane were a few observations. Failure initiated gradually with small movement on slope and then extended along weaker portions. Mainly method of mining was either block caving or sub-level caving except Shirengou iron ore mine, presence of fault zones and mining near the toe of overall pit slope where barrier thickness was insufficient were identified as the causes of instability. Impact of underground blasting is also visualized. In case of caving method, it is obvious to have instabilities if other factors are not given proper attention.

The crown pillar faces a new challenge in dealing with the change in geo-mining conditions from open pit to underground. The vertical stresses acting on the pillar usually increases with increasing depth of mining, which is not commonly encountered in the case of surface crown. Hence the behavior of the pillar under these stresses must also be thoroughly examined for optimum design of crown, and if necessary, an additional support system must be provided to ensure the pillar's stability. Critical factors must also be identified to have a proper understanding of the mining scenario and finally to have an optimum design methodology.

4. Indian geo-mining condition

Compared to coal mines, limited number of metal mines exists in India. Orebody width varies from 1 m to 75 m in the range of 5, 12, 20 and 35 m to 75 m). At few places, it is in range of 80–120 m. Orebody dip is in the range of 55 to 85°. Ultimate Pit depth varies from as shallow as 20 m to as deep as 380 m. Number of small mines in and around Central India has undergone transition at a very shallow pit bottom depth of 20–80 m mainly due to small lease area and limited lateral extent. Strike length varies from merely 100 m to as long as 2200 m. RMR varies from Fair to Good in case of orebody but for hangwall and footwall poor RMR is also observed. In Sukinda chromite mine however ore is very soft up to a certain depth and as of today no suitable technique is evolved to extract the soft ore deposit. In Rampura Agucha mine shear zone exist in orebody and a crown thickness of 70 m is kept to ensure its stability. List of the metal mines with pit bottom depth is given in **Table 1**.

From the analysis of literature and experience of modeling stability in different case studies various parameters that influence crown pillar stability are identified as controllable or uncontrollable factors as given in **Table 2**.

Variant of sub-level stoping like Long hole stoping or Blast hole stoping and Cut and fill method of mining is more commonly in use at Indian Mines. Top down sequence of mining has gain popularity worldwide. Similar trend has started in India too. In this method crown is subjected to load at early stage of underground mining due to drill drivage being made below it whereas in case of bottom up it is put under load at later stage of mining. Hence good strength backfill is desirable to bear the responsibility to isolate different working areas. Nowadays use of paste fill has gained importance as it helps in reducing tailing waste and providing sufficient strength to wall rocks, stope back and stope brow after filling. However, for very small mines its use cannot be economical rather a group of mines in close proximity can thought of paste use from a single plant. Vertical extensiveness of deposit does matter. Small mines having limited depth of working like 60 m – 100 m below open pit are usually adopting bottom-up sequence of mining and using bottom ash from thermal plants as fill material as sand is becoming scarce with time.

In most of the mines being small in extent measured in situ stress values are not available. Big mines does opt for such measurement in their mine. In rare cases, a few pit had been backfilled. Such fill offers confinement to the crown, but in rainy season water infiltration may cause backfill to offer dead load on crown and can lead to failure if high factor of safety is not considered. It will worsen further if rock mass is highly fractured. Hence, it is usually advisable to keep the open pit unfilled till the completion of underground workings and divert rain water to a deeper level sump or maintain a crown of at least 60 m as per Reg. 150 (3). Accessible benches and toe of the slope can be shotcreted to slow down rock surface deterioration. Detailed

Name of the Mine	Mineral extracted	Open pit depth (m)
Rampura Agucha	Lead-Zinc	380
Malanjkhand Copper Project	Copper	240
Sukinda Chromite Mine	Chromite	125–175
Palaspani Manganese Mine	Manganese	92–125
Dongri Bujurg Mine	Manganese	160 (Proposed)
Kandri Mine	Manganese	83–100 (approx.)
South Kaliapani	Chromite	115
Beldongri Mine	Manganese	50–55
Kathpal Chromite mine (closed)	Chromite	38
Pandharwani Manganese Mine	Manganese	50
Ramrama Manganese Mine	Manganese	22–25
Sarubil Chromite mine	Chromite	
Jagantola Manganese Mine	Manganese	36
Katangjhari Mine	Manganese	36
Jam mine	Manganese	15–42
Netra Mine	Manganese	16
Nuashahi Chromite mine (closed)	Chromite	30–35 (approx.) (pit backfilled)

Table 1.
List of the metal mines with pit bottom depth.

Controllable Factors	Un-Controllable Factors
Ultimate pit depth	Orebody geometry i.e., orebody width and dip,
Exposed width of crown at pit bottom	Rockmass quality i.e. rock mass strength, RMR
Extent of opencast working	Geological disturbances – fault or such hidden structure
overall slope angle	Presence of water
Effect of open pit blasting if simultaneous working exist	In situ stress state
Sequence of mining	stress state on account of open pit
Method of mining with Backfill	Rainwater influence/infiltration

Table 2.
Various parameters that affects crown pillar stability.

geotechnical mapping on regular basis is also highly essential at different stages of mining to be abreast of likely existence of discontinuity at sensitive places.

For design of this crown pillar no identified methodology for Indian mines is available. Statutory requirement of Directorate General of Mines Safety (DGMS) states that there should be a barrier of 60 m against water body/waterlogged workings or for two different workings approaching towards each other. Considering this regulation, 60 m crown had been usually left during transition phase of mining in some mines.

Whereas in a few small shallow depth mines 15–20 m crown is also left to isolate open pit from underground workings. Therefore, leaving 60 m against waterlogged workings or site-specific design by numerical modeling is followed by practicing engineers.

5. Case study

A mine with steeply dipping orebody exploiting manganese ore through opencast workings is considered for small scale analysis. Depth of working has reached 103 m from surface. Being a small lease deeper level orebody exploitation is planned by prevalent cut and fill underground workings. Width of orebody ranges from 2 to 7 m. The orebody is steeply dipping at 70–85° due south. The strike length of the deposit is about 1300 m in East-West direction. From field investigations, Immediate roof Rock Mass Rating (RMR) in orebody has been estimated as 54 which is "fair category" rock mass whereas RMR of hangwall and footwall rock mass has been estimated as 61 which is "good category" rock mass. Stopes were at 45 m level interval. Overall pit slope of 40°-50° exist at the site.

Two-dimensional plain strain models were solved for surface crown design. Simulation was performed by using FLAC3D software (developed by ITASCA Consulting Group, USA) based on explicit Finite Difference method. Plain strain is a special case of three-dimensional situation. A model of size 1650 m × 1 m × 1100 m is prepared with an element size of 1.0 m in the area of interest. Existence of open pit excavation after reaching ultimate pit limit is simulated with no backfill and no accumulation of water. Initial model is run in elastic state to generate virgin stresses. The model state is then changed to Mohr model and open pit excavation is made. Since crown should be stable till underground operations are over, full height stopes up to the crown is excavated in the model to visualize stability scenario.

Since measured horizontal in situ stresses are not available for this mine, equation for mean horizontal stresses as stated in Eq. (1) based on the theory of in situ stresses [19] is being used to estimate horizontal stresses.

$$\sigma hm = \frac{\beta EG}{(1-\vartheta)}(H+1000) + \left[\frac{\vartheta}{1-\vartheta}\right]\gamma H \tag{1}$$

where, σhm = mean virgin in situ horizontal stress in MPa.

H = depth in m.

E = modulus of elasticity of rock.

ϑ = Poisson's ratio of rock.

γ = Rock density/unit depth.

G = Thermal gradient.

β = Coefficient of thermal expansion.

Vertical stress was taken as the gravitational load due to self-weight of the overlying strata,

$$\sigma v = \gamma H \tag{2}$$

Intact Rock properties are shown in **Table 3** as follows.

S. N.	Physico-mechanical Property	Orebody	Wall rock
1.	Elastic Modulus (E) in GPa	13,000	13,000
2.	Poisson's ratio (ν)	0.25	0.25
3.	Density (Kg/m^3)	3000	2800
4.	Cohesion (KPa)	5500	5800
5.	Angle of Internal friction	35	38

Table 3.
Intact rock properties of various materials.

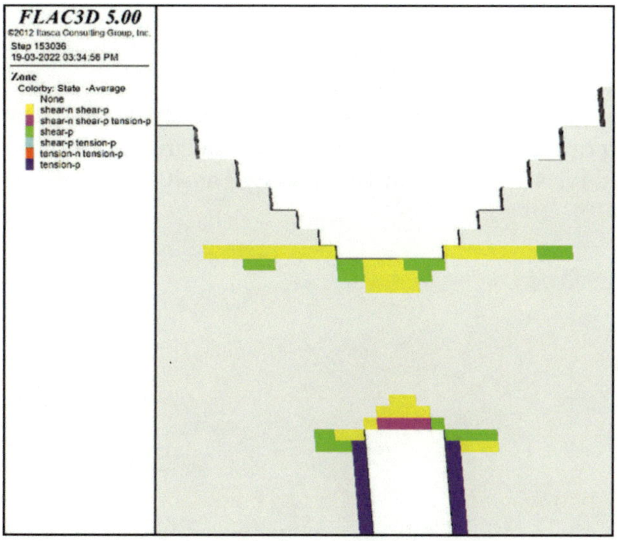

Figure 1.
View of stability of crown pillar for 7 m wide orebody.

Simulation was performed by using *FLAC3D* software (developed by ITASCA Consulting Group, USA) based on explicit Finite Difference method. For simulation stope length of 60 m is considered in analysis along with a 6 m wide rib pillar and 5 m height crown in between two levels. Level interval of 45 m is considered. Due to limited information about existence of ore beyond 200 m depth, only two-level extractions is considered. To simulate the existence of opencast workings or existing pit, overall pit slope of 40° and ultimate pit depth of 103 m from the surface is considered.

From the detailed analysis, 15 m crown is found to be stable for 7 m wide orebody. Crown pillar stability status for one particular section is shown in **Figure 1**. All the colored blocks that are in shear or tension are unstable zone. Failure in the backfilled stope back does not extend up to the open pit level failure zone in crown.

6. Parameter sensitivity analysis

Same model is extended for studying the effect of depth, different crown thickness, horizontal to vertical stress ratio and overall slope angle. For this comparison no backfill after stope excavation is considered. Different locations for monitoring the displacement are considered as marked on the section (**Figure 2**).

Stress state in the crown is also analyzed. Stress ratio that is, ratio of horizontal to vertical stress of 0.63 and 0.97 is used to compare status of crown pillar. In Indian mines, open pit depth varies from 50 to 400 m. Hence, two extremes of 103 and 400 m is studied. When depth of pit is increased from 103 to 400 m and stress ratio is kept constant at 0.97, 15 m crown fails. Stress state for 15 m crown at 103 and 400 m depth is shown in **Figure 3**. Stress concentration in the core of crown pillar has almost doubled from the initial values to about 30 MPa at 103 m depth. High stress concentration which was experienced in corners of crown pillar at 103 m depth is now visible at pillar core in crown at 400 m depth.

When crown thickness for this case is increased to 40 m stress state becomes moderate for 103 m depth but is still high for 400 m depth although not failed as shown in **Figure 4**. For 40 m crown at 103 m depth stress concentration at core is 15 MPa and it increases to about 30 MPa at 400 m depth.

High stress concentration is observed near the corners at pit bottom and zone in contact with underground excavation. Hence displacement at monitoring points 3, 4, 5, 7, 8 is compared for 40 m crown at 103 m as well as 400 m depth and is presented

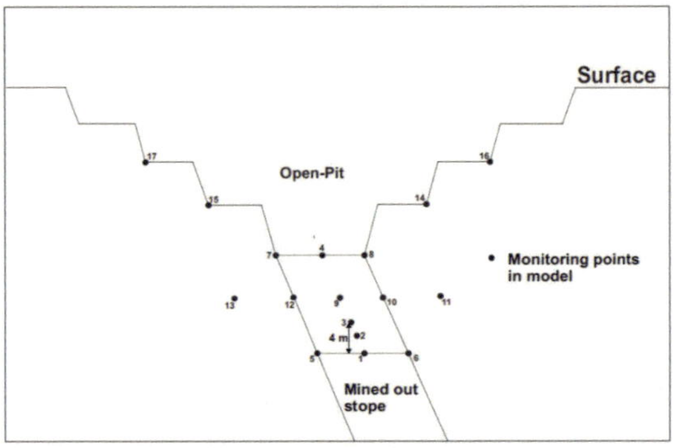

Figure 2.
Monitoring points location marked on cross-section.

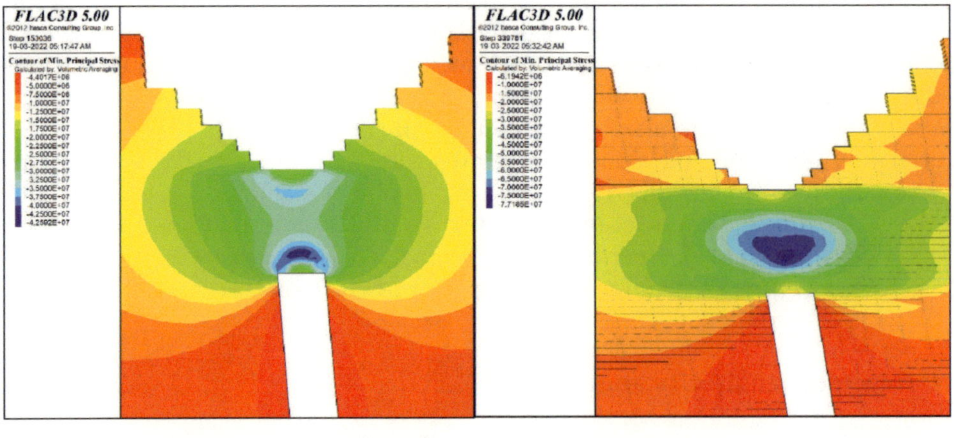

Figure 3.
Stress state of 15 m crown at UPD of 103 m and 400 m respectively for overall slope angle of 40°.

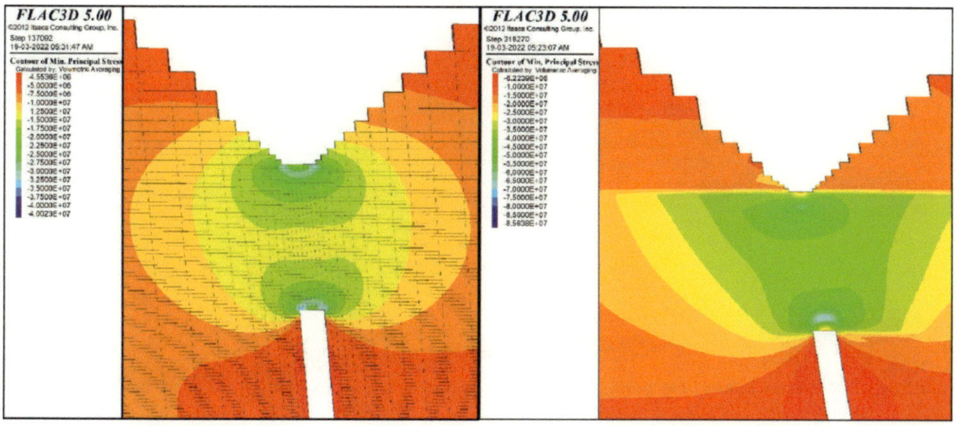

Figure 4.
Stress state of 40 m crown at UPD of 103 m and 400 m respectively for overall slope angle of 40°.

in **Figure 5**. Depth appears to be a major controlling factor as high displacement is observed at all the monitoring points when depth of open pit is 400 m and as compared to 103 m depth.

Similarly, displacement at monitoring points 3, 4, 5, 7, 8, 14 & 15 for 40 m crown at 103 m depth and different stress ratio of 0.63 and 0.97 is compared and is presented in **Figure 6**. At low stress ratio tensile movement is visible due to reduction in lateral confinement whereas uplift is possible when stress ratio is more and on footwall side slope monitoring point it increases sharply mainly due to effect of dip of the orebody.

Displacement at slope monitoring points (14 & 15) for 40 m crown at 103 m depth and different overall slope angle of 40° and 50° is presented in **Figure 7** displacement in hangwall side remains almost same with change in overall slope angle while footwall side slope has underwent more movement when slope angle is 50°.

More detailed parametric study and analysis in terms of magnitude of displacement and stress state is under progress for wider range of crown thickness and other parameters. However, from obtained results depth of open pit, overall pit slope angle and stress ratio are found to have significant impact on crown pillar behavior and its design. Therefore, in addition to orebody geometry, RMR, height of stope, open pit geometry viz., overall slope angle, pit depth, influence of geological disturbances

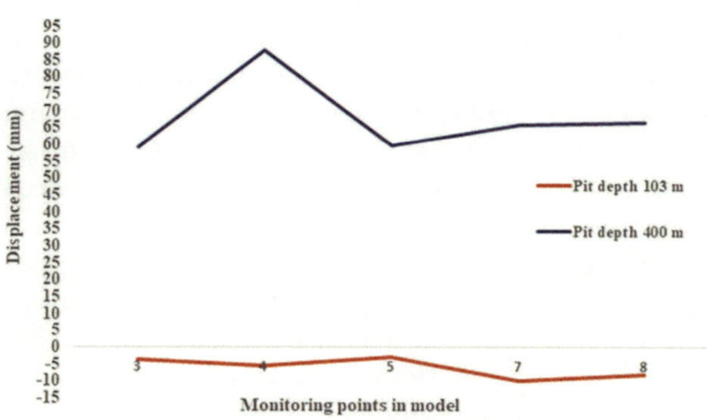

Figure 5.
Displacement at selected monitoring points for 40 m crown at different depths and stress ratio of 0.97.

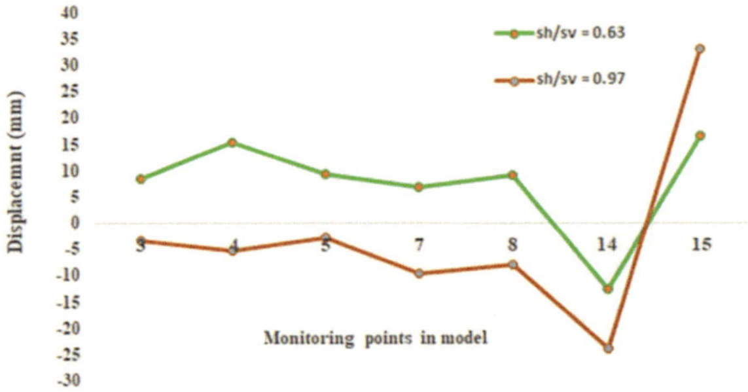

Figure 6.
Displacement at selected monitoring points for 40 m crown at 103 m depth and different stress ratio.

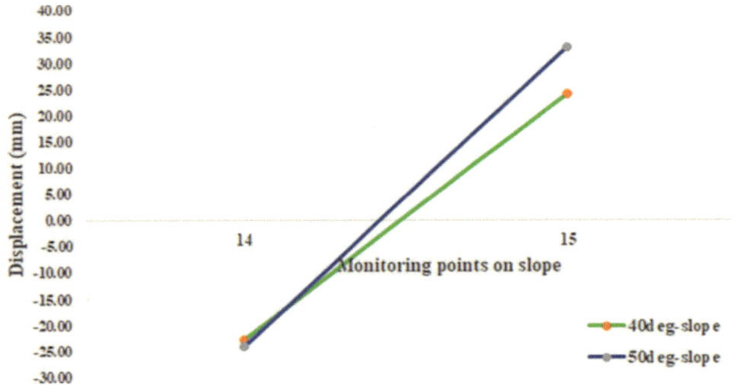

Figure 7.
Displacement at slope monitoring points (14 & 15) for 40 m crown at 103 m depth and different overall slope angle of 40° and 50°.

along with hydrogeological condition and in situ stress condition as well as reoriented stress on account of open pit mining should also be considered in design of a suitable thickness of crown pillar.

7. Conclusion

Determination of optimum thickness of crown pillar has been researched for many years but limited design guidelines are available. Initially, major concern was restricted to design of surface crown where underground working exists below surface infrastructure. Failure pattern for surface crown have also been analyzed. Although study was later extended to shallow dip stopes but influence of open pit working is probably not analyzed in detail. Although in literature, a few case studies are discussed where instability of crown pillar and open pit benches have occurred. Most of them were having caving method of mining below existing open pit. Hence method of mining and mining sequence also has a role in crown pillar design.

More detailed parametric study is going on but small-scale analysis is done to analyze the effect of depth of open pit, overall pit slope angle and horizontal to vertical stress ratio and are found to have significant impact on crown pillar behavior and its design. Therefore, in addition to orebody geometry, RMR, height of stope, open pit geometry viz., overall slope angle, pit depth, influence of geological disturbances along with hydrogeological condition and in situ stress condition as well as reoriented stress on account of open pit mining, mining method, sequence of mining should also be considered in design of a suitable thickness of crown pillar.

Various computer codes such as FLAC3D, UDEC, PFC, ANSYS, GEOSLOPE etc. are available to address specific mining issues. Data generated by empirical and observation techniques helps to validate the model and make the predictions more realistic. But obtaining data for each and every parameter is not always possible. Therefore, Carter (2014) has emphasized a coupled approach of numerical modeling and empirical approach for solution instead of using just one type of approach. A coupled approach of empirical, observational and analytical methods is generally beneficial.

Acknowledgements

The author is grateful to The Director, CSIR-CIMFR, Dhanbad for permission to publish this article. Views expressed are of authors and not necessarily of the institute to which they belong. This work is financially supported by Ministry of Mines. The authors gratefully acknowledge the support. We are thankful to mine management for providing the data of the mine.

Conflict of interest

The authors declare no conflict of interest.

Author details

Chandrani Prasad Verma* and Abhay K. Soni
CSIR-Central Institute of Mining and Fuel Research, Civil Lines, Nagpur, Maharashtra, India

*Address all correspondence to: chandranidp@gmail.com

References

[1] Hutchinson DJ. A review of crown pillar stability assessment and rehabilitation for mine closure planning. In: The Proceedings of 4th North American Rock Mechanics Symposium, Seattle, Washington, July 31 – Aug. 3, 2000. Rotterdam: A.A. Balkema; 2000. Paper No: ARMA-2000-0473. ISBN: 9058091554

[2] Kumar H, Deb D, Chakravarty D. Design of crown pillar thickness using finite element method and multivariate regression analysis. International Journal of Mining Science and Technology. 2017;**27**(6):955-964. DOI: 10.1016/j.ijmst.2017.06.017

[3] Shiguo S, Weidong L, Yujuan Z. The safe thickness of top plate in the process of the underground excavation transferred to opencast excavation. In: Proceedings of International Symposium on Energy Science and Chemical Engineering (ISESCE 2015). China: Atlantis Press; 2015. pp. 124-129

[4] Xu S, Suorineni FT, Li YH, Jin CY. Use of an artificial crown pillar in transition from open pit to underground mining. International Journal of Rock Mechanics and Mining Sciences. 2019;**117**:118-131. DOI: 10.1016/j.ijrmms.2019.03.028

[5] Carter TG. Design lessons from evaluation of old crown pillar failures. In: Proceedings of International Conference on Surface Crown Pillars for Active and Abandoned Metal Mines, Timmins, Canada. Ottawa, Canada: Energy, Mines, and Resources Canada; Canadian Govt. Pub. Center [Distributor]; 1989. pp. 177-187

[6] Carter TG. A new approach to surface crown pillar design. In: Proceedings of 16th Can. Rock Mechanics Symposium, Sudbury. Symposium from June 15-19. Sudbury, Ontario, Canada: Laurentian University; 1992. pp. 75-83

[7] Carter TG. Guidelines for Use of the Scaled Span Method for Surface Crown Pillar Stability Assessment. Ontario: Ministry of Northern Development and Mines; 2014. pp. 1-34

[8] Bakhtawar E, Oraee K, Shahriar K. Assessment of crown pillar thickness between open-pit and block cave mining. In: Proceedings of the 29th International Conference on Ground Control in Mining, Morgantown, West Virginia, USA. Morgantown, West Virginia: West Virginia University; 2010. pp. 325-332

[9] Xing-dong Z, Lian-chong L, Chunan T, Hong-xun Z. Stability of boundary pillars in transition from open pit to underground mining. Journal of Central South University. 2012;**19**:3256-3265. DOI: 10.1007/s11771-012-1402-x

[10] Xu N, Yanlin Z, Gang M. A numerical modeling method for predicting pit slope deformation and failure caused by underground mining under an open-pit. In: Proceedings of 13th International Congress of Rock Mechanics; Innovations in Applied and Theoretical Rock Mechanics. The Canadian Institute of Mining, Metallurgy & Petroleum and ISRM. Montreal, Quebec: ISRM; 2015. ISBN 978-1-926872-25-4; ISRM13 10-13 May 2015, Montreal, Canada;#20

[11] Kalenchuk K, Falmagne V, Gelover A, et al. Risk evaluation, design, implementation, instrumentation, and verification for crown pillar extraction at Pinos altos mine. Rock Mechanics and Rock Engineering. 2019;**52**:4997-5011. DOI: 10.1007/s00603-019-01801-z

[12] Nguyen PMV, Niedbalski E, Widzyk-Capehart Z. Determination of the crown pillar thickness between open pit and underground for coal mining. In: Proceedings of the 27th International Symposium on Mine Planning and Equipment Selection-MPES 2018. Switzerland AG: Springer Nature; 2019. pp. 393-404. DOI: 10.1007/978-3-319-99220-4_32

[13] Chen T, Mitri HS. Strategies for surface crown pillar design using numerical modelling – A case study. International Journal of Rock Mechanics and Mining Sciences. 2021;**138**:104599. DOI: 10.1016/j.ijrmms.2020.104599

[14] Golder Associates Report 881-1739 to CanMet on Crown Pillar Stability Back-Analysis. Report#23440-8-9074/01-SQ, CANMET

[15] Hutchinson DJ, Phillips C, Cascante G. Risk considerations for crown pillar stability assessment for mine closure planning. Geotechnical and Geological Engineering. 2002;**20**: 41-63

[16] Brummer RK, Li H, Moss A. The transition from open pit to underground mining: An unusual slope failure mechanism at Palabora. In: The South African Institute of Mining and Metallurgy International Symposium on Stability of Rock Slopes in Open Pit Mining and Civil Engineering. Cape Town: The South African Institute of Mining and Metallurgy; 3-6 April 2006. pp. 411-420

[17] Zhang P, Yang T, Yu Q, Xu T, Zhu W, Liu H, et al. Microseismicity induced by fault activation during the fracture process of a crown pillar. Rock Mechanics and Rock Engineering. 2015;**48**:1673-1682. DOI: 10.1007/s00603-014-0659-9

[18] Wang Z, Song G, Ding K. Study on the ground movement in an open-pit mine in the case of combined surface and underground mining. Advances in Materials Science and Engineering. 2020;**2020**:8728653. DOI: 10.1155/2020/8728653

[19] Sheorey PR. A theory for *in situ* stresses in isotropic and transversely isotropic rock. International Journal of Rock Mechanics and Mining Sciences & Geomechanics Abstracts. 1994;**31**(1):23-34. DOI: 10.1016/0148-9062(94)92312-4

Chapter 5

Eco-Energy Innovations in Revolutionizing Coal: Assessing Clean Technologies for a Greener Future

Manoshi Saikia, Samiran Bora and Prasenjit Saikia

Abstract

This chapter explores the potential for transformation brought about by eco-energy innovations within coal technology, investigating their capacity to mitigate environmental repercussions and propel the advancement of a more sustainable future. An array of clean coal technologies is scrutinized, encompassing pre-combustion methodologies like coal washing and gasification, combustion practices such as fluidized bed combustion and advanced pulverized coal combustion, as well as post-combustion remedies like carbon capture and storage (CCS) and flue gas desulfurization. The exploration extends to pioneering strategies for the sustainable utilization of coal, which includes co-firing alongside biomass, coal-to-liquids (CTL) and coal-to-gas (CTG) procedures, in addition to carbon utilization and conversion technologies. Moreover, it delves into the crucial significance of supportive regulations, financial encouragement, and market dynamics in propelling the adoption of these technologies. Through detailed case studies drawn from various regions including the United States, China, Europe, and India the chapter showcases successful implementations and underscores the global endeavors directed towards fostering cleaner coal practices. Ultimately, by embracing these eco-energy innovations, the chapter asserts that the energy sector stands to markedly diminish coal's environmental impact, thereby striking a balance between the imperatives of reliable energy provision and environmental stewardship.

Keywords: carbon capture, carbon utilization, coal-to-gas, coal-to-liquids, co-gasification, eco-energy, environmental repercussions, environmental stewardship, financial encouragements, flue gas desulfurization, fluidized bed, global endeavors, innovations, market dynamics, mitigate, pre-combustion, pulverized coal combustion, sustainable future

1. Introduction

In today's climate-conscious world, coal's historical role in energy is scrutinized due to its environmental impact. While coal has been essential for energy production, its carbon emissions and pollutants create a dilemma for citizens and policymakers seeking ways to meet global energy needs without harming the planet. This chapter critically examines clean coal technologies reshaping the industry, focusing on innovations from pre-combustion coal cleaning to advanced combustion methods and post-combustion carbon capture and storage (CCS). It also explores sustainable strategies like co-firing with biomass, coal-to-liquids (CTL) and coal-to-gas (CTG) processes, and emerging technologies that convert carbon emissions into valuable products, promoting a circular carbon economy.

This chapter expands beyond technological strategies, addressing how these innovations can be implemented and sustained through policies, financial incentives, and market factors. By examining case studies from the United States, China, Europe, and India, we explore global approaches to cleaner coal use. The goal is to find a balance between energy production and environmental conservation. This chapter discusses how ecological initiatives in coal technology could help achieve a more sustainable world.

2. Pre-combustion technologies

Coal purification before combustion enhances its economic value and environmental sustainability through various methods, including physical, chemical, and biological techniques. Physical purification, such as coal washing, effectively removes ash, pyritic sulphur, and trace elements like mercury, achieving up to 80% trace element removal. Drying methods, including solar drying and heat application, reduce coal's moisture content, improving transport efficiency and heating value. Blending different coal types and briquetting with suitable binders can reduce costs, stabilize toxic elements, and enhance combustion [1–3]. Coal cleaning, particularly through advanced gravity techniques like froth flotation, improves power plant efficiency, reducing SO_2 emissions by 40% and CO_2 by 5% [4–6].

3. Combustion technologies

Coal combustion through gasification is predominantly considered an effective and cleaner conversion process. Conventional combustion technologies around coal, such as fixed beds, fluidized beds, and entrained flow beds, are robust and have achieved large-scale industrial applications. The process of converting coal into product gas, including H_2, CO, CH_4, and H_2S, and other gases under high temperatures by using H_2O, air, or O_2 as the gasifying agents. The gases produced (syngas) were cooled down and subsequently purified to obtain the desired compositional gases. The syngas can augment to be used for effective production of electricity, fuel or other high-value-added chemicals like DMF, Furan, Furfural, wax, dimethyl ether, etc. As researchers endeavored, a series of pioneer technologies have been investigated, such as supercritical water gasification [7], plasma gasification, chemical looping gasification, and decoupling gasification [8]. The combustion technologies are guided by the rank of coal [9], the type of gasifiers and gasifying agents and the economics behind the end uses [10].

3.1 Fixed bed gasification

Fixed bed gasification is governed by a vertical reactor that has been further classified into two distinct reactor structures: updraft and downdraft. In updraft reactors, coal is feed from the top and moved down to the bottom while the gasifying agents are pumped from underneath in countercurrent flow. The syngas produced is extracted from the top of the reactor, and the residual ash is extracted from the bottom [11]. The gasification with updrafts reactors is more suitable for tar production with a high yield. While in downdraft gasifier, the gasifying agent is introduced in the middle part of the reactor in a concurrent flow, and the produced syngas are extracted from beneath the reactor [12]. For downdraft gasifiers, the carbon conversion rate is higher (90–95%) than any other with low tar yield and cleaner syngas. The gasification zones in both types are similar, with differences in their operability parameter and end results. There are significant disadvantages in both the types like limited feedstock particle size, low feeding rate and high sensitivity to slag formation.

3.2 Fluidized bed gasification (FBG)

Fluidized bed gasification represents a major advancement in thermal conversion technology, offering efficient and cleaner processes compared to other methods. Classified into types such as bubbling, spouted, circulating, and dual beds, it has evolved from technologies like Ende, U-Gas, and HTW. Unlike conventional reactors where coal is mixed with the bed medium from the start, in fluidized bed gasification, coal is introduced after reaching the target temperature (800–1000°C), ensuring stable fluidization with uniform particle size (1.00–0.6 mm). This method achieves rapid mass and heat transfer. Oxy-combustion in fluidized bed reactors is gaining attention, with Spain's CIUDEN 30 MWth circulating fluidized bed boiler showing significant reductions in NOx and SOx emissions [13]. Agglomerating fluidized bed gasification is also promising for producing synthetic natural gas from high ash and low-rank coal, as demonstrated in plants in Yunnan and Shanxi provinces. The lower carbon conversion rate, instability of the fluidized bed and high tar content in the produced syngas are major drawbacks of this gasifier type.

3.3 Entrained flow bed gasification

Entrained flow bed gasification gained momentum due to its capability of producing H_2 rich syngas with minimal amount of solid residue at relatively higher temperatures [14]. The feedstock and gasifying agent are injected through a special nozzle into the gasifier simultaneously from the top, and the produced gas is extracted from the mid of the reactor with a low discharge of ash from the bottom [14]. The operational temperature of this type of gasification lies around 1200–1700°C with relatively smaller particles, 0.001 cm, and lower residence time for a higher carbon conversion of coal. Companies like Texaco, Shell, Siemens, and others have induced and evolved with entrained flow bed gasification technologies [15].

Apart from the above-mentioned traditional methods, there are several modernized coal treatment technologies like Advanced Pulverized Coal Gasification, Supercritical Water Gasification (SCWG), Plasma Gasification, Chemical Looping Gasification and Decoupling Gasification that pave the way for cleaner, more efficient coal utilization, reducing the environmental footprint of coal.

3.4 Advanced pulverized coal gasification

Conventional coal-based power plants using stoke firing face challenges such as limited load fluctuation handling, difficulties in ash removal, and combustion interference due to ash. Advanced pulverized coal combustion systems address these issues by grinding coal into a fine powder, which is then introduced into the combustion chamber suspended in a gasifying agent like hot air. Supplementary air ensures complete combustion [16]. Recent studies on pulverized coal gasification in counterflow filtration combustion mode show high chemical efficiency by combining entrained flow and updraft-packed bed processes [17]. Additionally, the selective non-catalytic reduction process for pulverized coal boilers, optimized through computational fluid dynamics, offers a cost-effective solution with lower maintenance and capital costs, making it suitable for smaller units like waste incineration plants [17]. This advanced coal gasification process effectively addresses temperature window, residence time, and reagent injection efficiency in the reaction zone, which hitherto was limitations in anterior coal gasification processes, as discussed.

Advanced pulverized coal combustion is further classified according to consumption rate and efficiencies based on the atmosphere and output demand, as described in **Table 1**.

3.5 Supercritical water gasification

Supercritical Water Gasification (SCWG) is an advanced coal conversion technology that converts coal into hydrogen (H_2) and carbon monoxide (CO) in a supercritical water environment, eliminating NOx and SOx emissions [19]. Supercritical water, with its unique physicochemical properties like low density, high diffusion, and weak hydrogen bonding, is the gasifying agent, enabling efficient coal conversion at 500–850°C and 24–29.7 MPa. SCWG offers several benefits, including lower gasification temperatures, high carbon conversion efficiency, and reduced CO_2 emissions [20, 21]. It also converts coal elements like sulfur, nitrogen, and phosphorous into inorganic salts, further reducing pollutants. A variation in feedstocks with syngas percentage is discussed in **Table 2** in the experimental scale. Despite its advantages, SCWG faces challenges such as incomplete gasification at lower temperatures, equipment corrosion, and limited large-scale implementation due to insufficient technical data. Nonetheless, SCWG holds promise for sustainable coal utilization in the near future [22, 24, 26–31].

3.6 Plasma gasification

Plasma gasification is an advanced coal conversion technology where coal reacts with gasification agents (H_2O, O_2, or CO_2) in an oxidizing arc plasma atmosphere to produce syngas containing CO, H_2, and CH_4. Operating at high temperatures (1200–1700°C), plasma gasification maximizes coal conversion efficiency (~100%) and

Technology	Efficiency (%)	Coal consumption (gm/kWh)
Sub-critical	<35	≥380
Supercritical	35–40	380–340
Ultra supercritical	40–45	340–320
Advanced Ultra supercritical	45–52	320–290

Table 1.
Classification of pulverized coal combustion plants efficiency and consumption [18].

Ref.	Feedstock	Temperature (°C)	Carbon gasification efficiency	Syngas content (%)	Method
[22]	Yimin lignite	700	—	59.86	Experimental
[23]	High-volatility bituminous coal	650	13	~72	Experimental
[8]	Chinese coal	600 ~ 750	—	67.6	Simulation
[24]	Hongliulin coal	800	—	67.06 ~ 71.3	Simulation
[25]	Coal and sorghum	500	—	49.2	Experimental

Table 2.
Comparative table of different supercritical water gasification technologies [10].

reduces tar production. The plasma source provides the heat required for endothermic gasification reactions. It transforms organic waste into purified syngas, converts toxic residues into glassy materials, and increases H_2-rich syngas production. Plasma gasification is classified into oxygen, CO_2, steam, and plasma torch-assisted types. Despite its advantages, challenges remain, including the need to cool hot syngas, the high cost and impurities of catalysts, technical issues like high energy demand and short electrode life, and concerns about hydrogen production costs, commercial viability, and CO_2 emissions. The prevailing research is on optimizing plasma generators, gasification agents, feeding rates, and other parameters to address these issues and enhance the technology's potential [32–34]. A comparison of different feedstocks and gasifying agents are highlighted in the **Table 3** with an experimental scale approach. Addressing these issues is crucial for its future optimization.

3.7 Chemical looping gasification

Chemical looping gasification is an advanced coal gasification technology that produces hydrogen-rich syngas with high purity, high heating value, and low tar content. It substitutes molecular oxygen with an oxygen carrier for partial coal oxidation [38]. The process involves two main reactors: a fuel reactor, where coal reacts with oxygen to produce syngas (CO, CO_2, H_2, and CH_4), and an air reactor, where the reduced oxygen carrier is re-oxidized. The exothermic reaction in the air reactor provides the necessary heat for the fuel reactor, enhancing overall efficiency. The metallic elements in the oxygen carrier also exhibit a catalytic effect, further optimizing the gasification process. **Table 4** highlights the effect on syngas yield of different coal feedstocks by oxygen carrier in different experimental scale. Operating in a reducing

Ref.	Feedstock	Gasifying agent	Temperature (°C)	SynGas yield (%)	Method
[35]	Municipal solid waste and coal	—	2500	43.73	Simulation
[36]	High ash bituminous coal	Steam	—	96.4	Experimental
[34]	Biomass and coal	Air/steam/O_2	2000	~30	Simulation
[37]	Lignite and used car tyres	Steam and air	1226.85	73.81	Simulation

Table 3.
Comparative table of different Plasma gasification technologies [10].

Ref.	Feedstock	Gasifying agent	Oxygen carrier	SynGas yield (%)	Method
[39]	Lignite coal	Steam	$NiFe_2O_4$	77.9	Experimental
[40]	Zhundong lignite coal	Steam	Cu-Fe-Mg	~20	Experimental
[41]	Meihuajing coal	Steam	CuO	57.68	Experimental
[42]	Bituminous coal	Steam	CuO	77.186	Simulation
[43]	Ningdong coal	Steam	Fe_2O_3	70	Experimental
[44]	Lignite coal	Steam	$CuFe_2O_4$	89.55	Experimental

Table 4.
Comparative table of different chemical looping gasification technologies [10].

atmosphere, this technology significantly reduces SOx and NOx formation. Chemical looping gasification is superior to other methods in reducing pollutant emissions, tar generation, and the need for oxygen separation, making it a promising approach for hydrogen production from coal. Extensive researches underway to optimize catalysts, oxygen carriers, and operational parameters [39, 45, 46].

There are considerable amounts of challenges associated with chemical looping gasification technologies like:

- Efficient design of oxygen carriers with high activity, anti-carbon deposition, high selectivity, better cyclic stability, etc. [47].
- Reaction performance of the oxygen carriers, such as ash content, pollutant gases, solid residues, etc., will also cause a significant impact [48, 49].
- Optimum design for gasification reactor [50].
- System integration of chemical looping gasification like reaction and product purification, process optimization, etc. [41].
- Lack of experience in implementation of large-scale application, in order words, engineering constrains [43].

3.8 Decoupling gasification

In conventional gasification, multiple intertwined reactions occur in a single reactor, making process optimization challenging. Decoupling gasification offers an advanced alternative, where coal is decomposed at mild temperatures to separate volatiles, tar, and char. The tar can be upgraded into liquid fuels or chemicals and gasified to produce hydrogen-rich syngas, which can be used for chemical synthesis, electricity generation, or hydrogen production [51]. **Figure 1** represents the schematic diagram of decoupling conversion of coal to various volatiles and liquid fuels.

Polygeneration *via* decoupling gasification offers efficient, clean coal utilization with high-quality syngas and versatile applications. With significant advantages in total coal utilization, there exists substantial number of challenges that need further research, such as

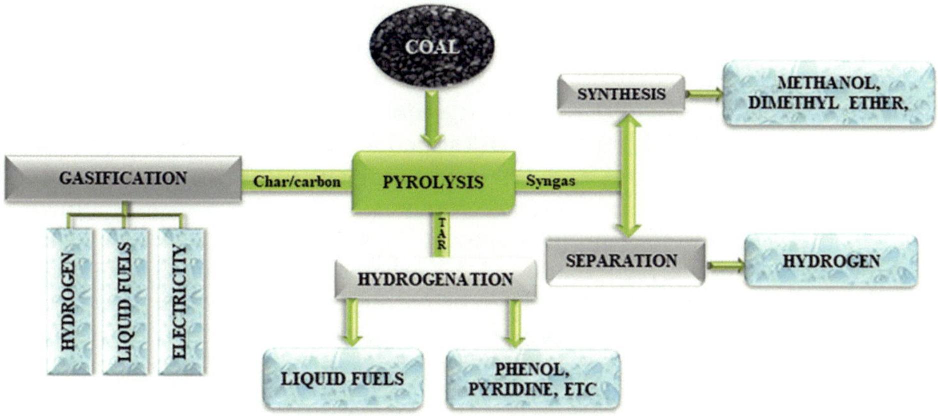

Figure 1.
Schematic diagram of coal decoupling conversion utilization.

- Key congestion in technologies in coal pyrolysis process-efficient pyrolysis furnaces, oil and gas dust removal of higher temperature, high-value-added products and environmental emission [52].
- Coal char gasification's lower reactivity requires selective agents, optimal conditions, and catalysts for industrial use [53].
- Polygeneration process of decoupling gasification is sophisticated as compared to other gasification, and it needs harmonizing couple between coal pyrolysis and char gasification (**Table 5**) [54].

4. Post-combustion technologies

The reduction in carbon emission is imperative, and post-combustion technologies are selective measures to achieve the objective. However, diverse processes are employed for the aforesaid, but in this section, the discussion will be on carbon capture and storage (CCU) and flue gas desulfurization (FGD).

Ref.	Feedstock	Temperature (°C)	Gasifying agent	SynGas content (%)	H_2 content (%)	Method
[54]	Coal Coke shenmu	1000	Steam	85.8	35	Simulation
[55]	Coal coke	900	—	96	56	Simulation
	Coke	1100	CO_2	64	2.8	Simulation
[56]	Bituminous coal	900	Steam	~40	~22	Simulation
[52]	Coke	1000	Steam	85.82	51.48	Simulation
[57]	Yunnan coal char	850	Steam	~80	~46	Experimental

Table 5.
Comparative table of hydrogen production from decoupling gasification technologies [13].

4.1 Carbon capture and storage (CCS)

Carbon Capture and Storage (CCS) is a key decarbonization strategy for coal-intensive industries aligned with the Paris Agreement. CCS can be applied in three stages: pre-combustion (CO_2 separated and stored, H_2 used for power generation), post-combustion (CO_2 removal from flue gases using chemical absorption), and oxy-fuel (CO_2 concentrated for storage *via* combustion in an O_2-rich environment). While oxy-fuel is highly efficient, its economic feasibility is limited by costs related to air separation and flue gas recirculation [57]. Carbon capture storage and utilization of coal for methanol production have been prime objectives for its sustainability, and Europe has been leading towards net carbon neutrality with CCS technologies [58].

4.2 Flue gas desulfurization (FGD)

Flue gas desulfurization is a technique based on scrubbing which uses alkaline reagents like Na or Ca-based to remove SO_2 from flue gas. Post-combustion flue gas desulfurization (FGD) involves wet processes generating stable sulfite FGD ($CaSO_3$) or sulphate ($CaSO_4$) FGD gypsum and dry processes consisting of spray dryer (SD), dry sorbent injection, etc. FGD effectively removes more than 90% of SO_2 in flue gases. Both wet and dry processes are further differentiated into better and more efficient technologies based on the requirements and economics of the total process function. The value-added products of gasification and FGD are materials rich with sulfite and have suitable potential regarding their applications in structural fills, stabilized road bases, agricultural applications, cement manufacturing, waste stream pollutant fixation, acid mine drainage mitigation, etc. In United States, sulfite and FGD gypsum and dry FGD are used for structural construction, mitigating the hazards of emissions and exploration of coal [59]; in Europe [60] and in China [61], manufacturing of cement and agricultural applications are prevalent.

5. Innovations in coal utilization

Sustainable utilization of coal has been an intriguing area of interest for researchers as well as commercial entities. Thermal energy for power generation has been the prime objective of coal conversion; however, conversion into chemicals and other carbon-rich substrates is the next stage from partial to total utilization of coal. In this section, predominantly on coal liquefaction and different upgraded gasification technologies alongside their evolution have been discussed comprehensively. **Figure 2** represents the schematic of various innovative methods developed for coal utilization.

5.1 Coal to gas (CTG)

In the late eighteenth century, William Murdoch introduced illuminating gas from coal, revolutionizing heating and cooking. By the mid-nineteenth century, advancements in converting lower-grade coal into clean, flammable gases like syngas were developed, leading to the widespread use of coal gas for urban and rural heating and lighting. Syngas is produced *via* coal gasification, a thermochemical process where coal reacts with an oxidizing agent. The yield and quality of syngas depend on coal rank, contents, and gasifier type. Key gasification technologies include fixed bed (Lurgi, BGL), Fluidized Bed (HTW, U-gas, KRW, and KBR), and entrained bed

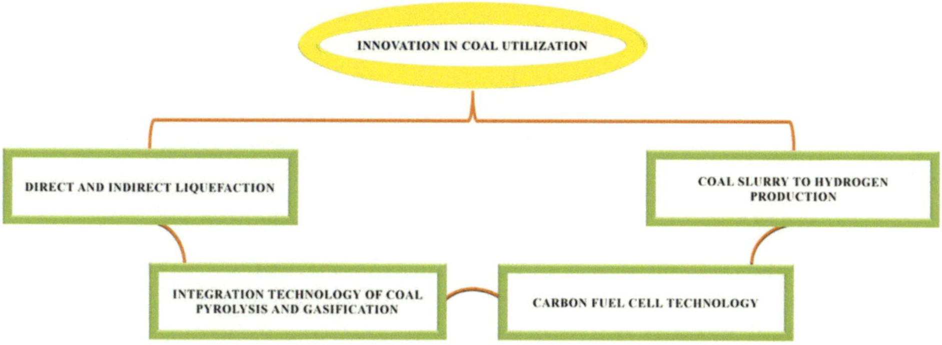

Figure 2.
A schematic representation of the innovation in coal utilization.

(GE, Shell, HT-L, and EAGLE). Fixed bed downdraft gasifiers offer easy operation, low tar production, and suitability for high-ash feedstock [62, 63]. Slag formation in entrained bed gasifiers handling high-ash coal reduces performance. Air as a gasifying agent introduces 50–60% nitrogen in syngas, aiding purification and eliminating the need for air separation and recycling compressors [64]. A combination of coal-steam gasification of higher efficiency and a one-step methanation process for syngas production is proposed in Ref. [65], where a regenerative unit recovers the sensible heat of the syngas through oxidant steam and then converts into chemical energy of syngas thereby increasing the cold gas efficiency 8.8% higher than other gasification. The one-step methanation process eliminates the water-gas shift. PGCC-Oxy, an advanced IGCC variant, uses O_2 and flue gas for combustion, offering higher thermal efficiency but lower coal carbon utilization [66].

Recent advancements in coal char-CO_2 gasification enhance H_2-rich syngas production and coal utilization. Decoupling gasification, driven by the char-CO_2 reaction, shows promise with low irreversible losses and complete coal utilization [67–69]. The performance of the char-CO_2 gasification is primarily dependent on coal types, catalysts and operational parameters.

- The low-ash sub-bituminous coal showed higher cold gas efficiency under microwave gasification [70]. The Ca-induced coal char exhibited superior gasification performance than Si and Al-induced char [71].

- The mass and heat transfer during char-CO_2 gasification is proportional to pressure, and there are researches attributing to the various differentials for maximizing output [72, 73].

- Adding catalysts like sodium to coal lowers activation energy in CO_2 gasification, enhancing carbon oxidation and CO formation. These catalysts significantly boost reaction rates and syngas production [74–78].

5.2 Coal-to-liquids (CTL)

The concept of coal-to-liquids is promising, with immense output to mitigate lower-grade coals. As discussed in Section 2.1 on syngas, Coal-to-liquids (CTL) technology, particularly through indirect coal liquefaction, converts coal into liquid fuels like diesel, gasoline, and jet fuel. This process, based on syngas production and Fischer-Tropsch (FT) synthesis, also produces valuable by-products like motor oils and paraffin waxes. Historically, CTL was pivotal during WWII, with Germany producing synthetic fuel, and since the 1960s, South Africa has consistently utilized CTL. This technology offers a promising solution for addressing global oil demand and promoting cleaner coal utilization. The Bergius process (1919) converted coal to liquid fuels *via* hydrogenation, crucial during WWII in Germany [79]. The Karrick process (1920) refined coal using pyrolysis to produce fuels. Post-WWII, the US attempted to replicate German processes, leading to SASOL's CTL plants in South Africa, which was crucial for its energy economy [80].

CTL technology involves two main processes: direct and indirect coal liquefaction. This converts coal into liquid hydrocarbons or syngas, which are then refined into liquid fuels. Additional methods include pyrolysis and methanol-to-gasoline conversion. **Figure 3** represents a flowchart showing various steps involved in the CTL technology with both direct and indirect liquefaction methods.

5.2.1 Direct coal liquefaction (DCL)

The Bergius process forms the basis of DCL, directly converting coal to liquid hydrocarbons *via* pyrolysis/carbonization and hydrogenation.

It splits coal into shorter hydrocarbons by adding hydrogen under high pressure and temperature Eq. (1), thus eliminating an intermediate gaseous stage as IDL process:

$$nC+(n+1)H_2 \rightarrow C_nH_{2n+2} \tag{1}$$

Direct coal liquefaction (DCL) processes, particularly the Bergius process, are effective for bituminous and sub-bituminous coals with high volatile

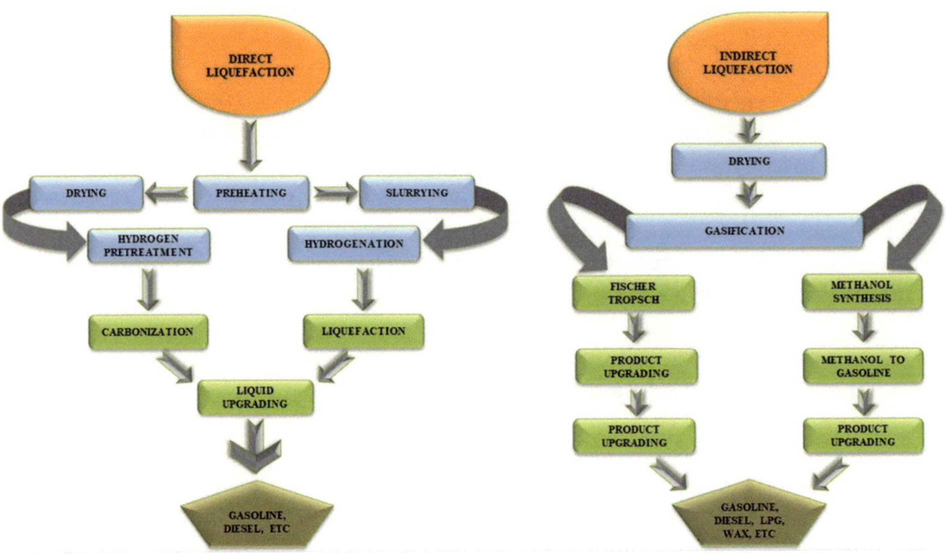

Figure 3.
Flowchart representation of coal to liquid process.

content, producing high-grade liquids and tar. Pyrolysis, with fast, flash, and slow variants, transforms coal into liquids and tars, with a maximum tar yield of 32% when methane reforming is integrated. Hydrogenation, or hydrocracking, uses high pressure, temperature, and hydrogen to convert coal into high-quality liquids with yields exceeding 70%. The first commercial DCL plant was built by Shenhua Group in China in 2002, with technologies like IGOR, HTI, FFI, and NEDOL used worldwide (**Table 6**) [81–83].

5.2.2 Indirect coal liquefaction (IDL)

Unlike direct coal liquefaction, indirect liquefaction uses gasification to break down coal into syngas, tar-like liquids, and char. The syngas is desulfurized and adjusted for hydrogen and carbon monoxide balance. Indirect coal liquefaction is cleaner and more self-sufficient, removing pollutants during gasification. However, it is more energy and capital-intensive compared to direct liquefaction [84]. Key industrial IDL processes include Chevron coal liquefaction, VEBA-combi-cracking, and solvent-refined coal [85]. VEBA-combi-cracking (VCC), developed by VEBA OEL (now BP), processes feedstock and additives in a slurry reactor under high pressure with H_2 gas. The unconverted material is separated, and light products undergo further H_2 treatment in a fixed-bed catalytic vessel. Originally for refinery residues, VCC can also process coal and plastics, demonstrated in the 1980s in Bottrop,

Country	Process	Reactor	Catalyst	Initiation
United States	Solvent Refined Coal-1 (SRC-1)	Coal slurry dissolver	—	1974
	Solvent Refined Coal-11 (SRC-11)	Coal slurry dissolver	—	1974–1984
	EDS	Entrained bed	Ni/Mo (Recycled)	1979–1983
	H-Coal	Fluidized Bed	Co-Mo/Al_2O_3	1979–1982
	CTSL	Fluidized Bed	Ni/Mo	1985–1992
	HTI	Suspended bed	GelCat	1990s
Germany	IGOR	Fixed bed	Red-mud (first stage), Ni-Mo/Al_2O_3 (second stage)	1981–1987
	PYROSOL	Countercurrent	—	1977–1987
Japan	BCI	Fixed Bed (two stage)	Fe-based (first stage), Ni-Mo/Al_2O_3 (second stage)	1986–1990
	NEDOL	Fluidized bed	Nature pyrite	1996–1998
United Kingdom	LSE	Stirred tank-type (first stage), Fluidized bed (second stage)	—	1983–1985
USSR	CT-5(FFI)	—	Mo	1986–1990
China	Shenhua 1	Suspended bed	Fe-based	2002
	Shenhua 11	Suspended bed	Fe-based	2004

Table 6.
Summarized history of direct coal liquefaction technologies worldwide [81].

Germany [86]. The solvent refined coal process (SRC-I) upgraded coal to produce cleaner fuel with lower ash and sulfur. It involved hydrogenation at high temperature and pressure. SRC-II, a modification, produced all-distillate low-sulfur fuel oils like naphtha, gasoline, and diesel. The process was eventually replaced by more advanced MTG technologies [87].

The recent most reliable advancements in CTL are two processes with which gasification is compatible; Fischer-Tropsch synthesis and ExxonMobil's methanol to gasoline (MTG). FT synthesis approach is based on Eq. (2):

$$(2n+1)H_2 + nCO \rightarrow nH_2O + C_nH_{2n+2} \quad (2)$$

In Fischer-Tropsch synthesis, a reaction between H_2 and CO (this mixture is generally called syngas) occurs through condensation and a suitable catalyst, resulting in the formation of numerous lengths of hydrocarbon chains. FT-process yields two products, namely based on Eq. (3) and Eq. (4).

$$nCO + 2nH_2 \rightarrow nH_2O + C_nH_{2n}\,(\text{Olefins}) \quad (3)$$

$$nCO + (2n+1)H_2 \rightarrow nH_2O + C_nH_{2n+2}\,(\text{Paraffin}) \quad (4)$$

FT synthesis products can be upgraded into jet fuel, gasoline, and naphtha, with high melting point wax and diesel. Indirect coal liquefaction is cleaner and more self-sufficient, removing pollutants during gasification. However, it is more energy and capital-intensive compared to direct liquefaction [84].

The products of FT-process can be upgraded into jet fuel, gasoline, and naphtha, with high melting point wax and high-yielding diesel. Developed in the 1920s, it is the primary method for producing liquid. It includes high-temperature processes for gasoline and low-temperature processes for diesel. Fixed and fluidized bed reactors are used, with Fe and Co-based catalysts being more practical than Ru and Ni.

Recent developments in methanol synthesis are catalysts, such as $Cu/ZnO/Al_2O_3$ (CZA), have improved efficiency. Despite FT synthesis's reliability and high yield, its

Shenhua Baotou Charcoal Chemical Co., Ltd	2011	GE gasification	Davy methanol synthesis
Shenhua Ningxia Coal Industry Group	2011	GE gasification	Lurgi methanol synthesis
Datang Inner Mongolia Duolun Coal Chemical Co., Ltd	2013	Shell gasification	Lurgi methanol synthesis
Baofeng Ningxia Energy Group Co., Ltd	2014	HT-L gasification	Davy methanol synthesis
Shaanxi Pucheng Clean Energy Chemical Co., Ltd	2015	GE gasification	Davy methanol synthesis
Shenhua Xinjiang Energy Co., Ltd	2016	Gas gasification technology	Davy methanol synthesis

Table 7.
Mega-methanol units in China with methanol capacity over one million metric ton/a [88].

break-even cost makes it economically unfeasible for large-scale commercialization, with only about 50% thermal efficiency in CTL processes (**Table** 7) [63, 88–91].

5.3 Co-firing and biomass blending

Co-firing biomass with coal enhances thermal efficiency and reduces CO_2 emissions. Biomass components like cellulose and lignin improve ignition and gasification rates [92]. Co-gasification of semi-coke and biochar shows both negative and positive synergetic effects at different carbon conversion stages. Blending walnut shells with coal improves gasification reactivity, and a 20% biomass blend with CO_2 capture reduces emissions by 144.76 kg/MWh, outperforming non-blended systems [93].

Increasing biomass ratios reduce pollutants like H_2S and HCN, with low-ash coal and high K/Na biomass showing optimal synergy. Modelling studies, like Aspen Plus, assess co-gasification parameters, while co-firing rice and corn husks with coal reduces SO2 emissions [94–99].

5.4 Carbon utilization and conversion

The coal chemical industry emerged in the late eighteenth century, establishing a comprehensive framework by the nineteenth century. The chemical industry transitioned from agricultural resources to coal in the early 20th century but lost its prominence during post-World War II with the rise of more efficient and versatile petrochemicals [100]. However, the twenty-first century has revived interest in coal, driven by escalating oil prices and its extensive reserves, emphasizing its utility in producing fuels and chemicals through processes like coking [101]. Despite its contributions, the traditional coal chemical industry faces significant challenges, including overcapacity, technical limitations in deep processing, high energy consumption, and serious pollution issues that impact the environment and public health [102]. **Figure 4** depicts the conversion of coal to various high-valued products (like carbon quantum dots, carbon nanosheets, nanotubes, etc.) as well as greener fuels (like higher alkanes-olefin, alcohols, etc.).

Advanced coal chemistry technology emphasizes pioneering approaches for the sustainable and effective use of coal resources. The contemporary coal chemical sector aspires to generate clean energy and chemicals by employing sophisticated coal

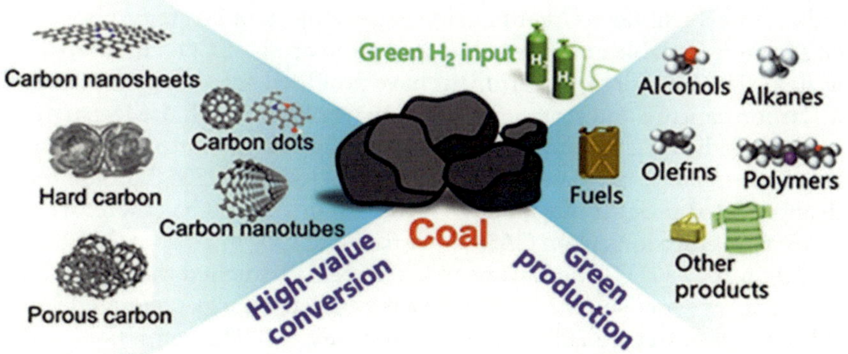

Figure 4.
Advanced carbon nanomaterials from coal. Reprinted with permission from Ref. [103].

conversion methodologies. This strategy aims to maximize coal resource utilization while reducing environmental repercussions.

Synthesis of Green Hydrogen: An innovative approach highlights the integration of coal conversion methods with green hydrogen production. This synergy enhances the creation of high-value outputs with a favorable hydrogen-to-carbon (H/C) ratio, while also striving to significantly lower CO_2 emissions [104].

Production of Olefins and Methanol: Olefins, including ethylene and propylene, are vital to the chemical sector, representing over 75% of downstream derivatives. Methanol acts as a fundamental precursor for various organic syntheses, encompassing the creation of olefins, formaldehyde, and acetic acid. These processes are crucial to produce everyday items like plastics and textiles. Methanol to olefins (MTO) from coal involves converting coal into synthesis gas, which is then processed into methanol. The methanol is catalytically transformed into light olefins like ethylene and propylene [105].

Production of Carbon Materials: Coal is acknowledged as an economical feedstock for generating high-value carbon materials such as Carbon Dots [106], Carbon nanotubes [107], Carbon nanosheets [108], and Porous Carbon for adsorption applications [109], as well as Hard Carbon anodic materials for batteries [110]. This viewpoint unveils new possibilities for coal utilization across diverse applications beyond conventional fuels and chemicals.

The development, methodologies, and challenges associated with both basic and advanced coal chemistry technology are examined, underscoring its significance in the chemical industry and the necessity for sustainable practices in the future. There is an urgent requirement for advanced technologies to promote the sustainable and eco-friendly use of coal resources, with a focus on diminishing CO_2 emissions and enhancing product diversity and quality.

6. Policy and economic considerations

Global regulatory energy policies follow the sustainable development goals (SDGs) of United Nations Framework Convention on Climate Change (UNFCCC) and other environmental convention. Governments and leaders are in commitment to achieve the equilibrium between life and lifestyle.

6.1 Regulatory frameworks

Canada's Long-Term Low-Greenhouse Gas Development Strategy, aims to produce electricity from renewable energy sources and pioneering of CO_2 emission devaluation by 2030. The US federal government have provided subsidies for the development of Carbon capture and storage technology wholly through the Department of Energy; the subsidies are tax provisions reducing the cost of capturing and storing CO_2. Since 2011 to 2023, lawmakers have allocated a total of US$5.3 billion for CCS research and related programmes. The government of Canada has therefore raised the price for CO_2 emissions from CA$ 10 per tonne of CO_2 in 2018 to CA$ 50 per tonne of CO_2 in 2022 [111]. The European Union (EU) launched the 2020 Climate and Energy Package focusing to enact a 20% increase in energy generated from renewable energy sources, a 20% cut of greenhouse gas (GHG) emission, and a 20% increase in energy efficiency (EC, 2009), however EU revised the package to 27% in 2015 [112]. In Asia, China and south Asian countries are surging in consumption of energy for their growth; therefore, a balance policy may jeopardize their nations' growth. Carbon credits and emission cap are incentives that are developed to mitigate emission from developed countries and provide growth to developing countries. In India, National Coal Gasification Mission has been drafted to liquefy and gasify 100 million tonnes of coal by the end of 2030 and to achieve this target ministry of coal has offered 50% rebate on revenue share for commercial coal block auctions if 10% of the coal produced is utilized for gasification [113].

6.2 Market dynamics

The global market size for coal liquefaction was valued at US$ 4.81 billion in 2021 and is projected to achieve US$ 7.01 billion by 2030, growing at a CAGR (compound annual growth) of 4.4% from 2022 to 2030 [114]. Asia-Pacific dominates an enormous share of the CTL market, with over 39% in 2021. Indirect liquefaction dominates at 58.7% due to technological advancements. The majority of CTL plants are located in China, with some in India (in its initial stages) and Indonesia. The abundant reserves of the regions and growing supply chain are expected to act as a catapult to its rapid growth. North America, currently building a robust production and supply network, is also expected to contribute significantly to the global market due to the high demand for clean fuel. Similarly, the Global market on Carbon Capture and Storage Utilization was valued at US$ 2.26 billion in 2022 and projected growth of US$ 2.52 billion in 2023 to US$ 6.16 billion till 2031 with a CAGR of 11.8% for the period (2024–2031) [115]. However, the challenge to achieve economic feasibility lies with the technologies being less sophisticated and diverse in application globally.

7. Case studies and global perspectives

Examining successful implementations of clean coal technologies around the world provides valuable insights into their potential and challenges. Case studies from across the globe offer critical insights into the potential and challenges of clean coal technologies. Successful implementations in various regions demonstrate the effectiveness of technologies such as carbon capture and storage (CCS), flue gas desulfurization, and coal-to-gas (CTG). These examples highlight how supportive regulatory frameworks, financial incentives, and public-private partnerships can drive the adoption of cleaner coal practices.

7.1 International

The US military's push for alternative jet fuels led to exploring coal liquefaction. Delays occurred as focus shifted to shale gas and oil. Nonetheless, research and projects continue under the US Department of Energy's National Energy Technology Laboratory, advancing coal-based fuel technologies (as discussed in Sections 2.1 and 2.2) [116]. Due to higher carbon footprints and greenhouse gas emissions, sustainable routes to liquefaction have received much attention. Active and functioning coal liquefaction projects are Adams Fork Energy- TransGas WV CTL developed by TransGas Development Systems (TGDS) in West Virginia, Future Fuels Kentucky CTL, Lima Energy Project by USA Synthetic Fuel Corp (USASF) in Ohio and US Fuel CTL by US Fuel Corporation. Carbon capture and storage are other avenues where focus was concentrated. Facilities that are actively operating in the United States are Lost Cabin (2013) in Wyoming, PCS Nitrogen (2013) in Louisiana,

Petra Nova (2017) in Texas, Illinois Industrial (2017) in Illinois and Red Trail Energy (2022) in North Dakota [91]. In the area of coal to chemicals, research is at nascent stage as the energy policies of the country are focused on other sustainable energy technologies and are in agreement with the SDGs. Similarly, within the Asian countries, China is leading the world as the largest methanol producer from coal through liquefaction and is rapidly expanding its capacity. The methanol produced in China is primarily converted to olefins (35.5%), methyl tert-butyl ether-MTBE (3.4%), MTG (3.8%), dimethyl ether (9.2%), methylamine (2%) and methane chloride (3.1%), oxidized to formaldehyde (24.4%), acetic acid (4.5%) and as fuel or additive to fuel (11.2%). Shenhua Xinjiang Energy Co. Ltd., Shaanxi Pucheng Clean Energy Chemical Co. Ltd. and Baofeng Ningia Energy Group Co. Ltd. are some mega-methanol units with over one million metric ton per annum [117]. Collaborations among institutions like the Dalian Institute of Chemical Physics and companies such as Sinopec have led to the development of advanced methanol-to-olefins (DMTO) technology. This sustainable route connects coal to methanol and subsequently to olefins, enhancing the efficiency of chemical production.

7.2 National

In India, a pilot and demonstration plant of CTL was set up in 2022 by Bharat Heavy Electrical Limited (BHEL) under the Clean Energy Research Initiative with a capacity of 0.25 metric tonne per day. India is presently the largest importer of methanol in the world as its production is about 1.5 million tonne per year and is produced from natural gas or naphtha. Methanol production from biomass has been a significant research focus, but in India, five key companies dominate methanol production: Gujarat Narmada Valley Fertilizers Company Ltd. (GNFC), Deepak Fertilizers, Assam Petrochemicals, Rashtriya Chemicals and Fertilizers Ltd. (RCF), and National Fertilizers Ltd. GNFC is one of the leading producers, along with Deepak Fertilizers and Assam Petrochemicals, all of which utilize technology from ICI for methanol production. In contrast, RCF and National Fertilizers Ltd. employ technology from Haldor Topsøe Ltd. [117]. Underground coal gasification plant in Gujarat, established by Gujarat Industries Power Company Ltd., is the first-ever pilot project in India, which is yet to be commercialized. As one of the largest countries, India has a huge potential for coal liquefaction and producing high-value chemicals, but the demand is exponentially higher than the production.

8. Conclusion

In conclusion, this chapter articulates the significant potential of eco-energy innovations within coal technology to substantially mitigate environmental repercussions while concurrently ensuring energy reliability. The exploration of eco-energy innovations in coal technologies demonstrates a clear pathway towards reducing the environmental impact of coal, even as it remains a vital energy source globally. The detailed examination of clean coal technologies—ranging from pre-combustion practices like coal washing and gasification, combustion techniques such as fluidized bed combustion, to post-combustion remedies including carbon capture and storage (CCS) and flue gas desulfurization—underscores their crucial role in mitigating the harmful emissions associated with coal use. Additionally, pioneering approaches like co-firing with biomass, coal-to-liquids (CTL) and coal-to-gas (CTG) technologies, along with carbon utilization and conversion, present promising avenues for the sustainable use of coal. However, while the technical potential of these innovations is evident, their widespread adoption is still hampered by financial, regulatory, and market challenges. Case studies from the National and International perspectives reveal that successful deployment of clean coal technologies is often tied to supportive regulatory frameworks, incentives, and robust policy enforcement.

Future research should focus on enhancing the efficiency of these technologies while simultaneously reducing their costs to make them more viable in developing markets. There is also significant room for advancing carbon utilization and conversion technologies, potentially transforming captured carbon into valuable products, which could help offset the costs of CCS and foster a circular economy. Moreover, integrating coal technologies with renewable energy sources, such as in hybrid systems, could offer a transitional pathway to a cleaner energy mix.

In conclusion, achieving a balance between the continued use of coal and environmental protection requires not only technological advancements but also robust political and economic support. Policymakers, industry leaders, and researchers must collaborate to create an environment conducive to innovation, ultimately driving the large-scale implementation of these eco-energy solutions. This combined effort is essential if the global community is to reduce coal's environmental footprint while ensuring energy security for the future.

9. Consent for publication

All authors of this manuscript have mutually agreed to publish and equally contributed in writing, framing and formulation.

Acknowledgements

The authors thank the Director of CSIR-NEIST, Jorhat, and the Head of the Coal & Energy Division for their unwavering support in conducting this research study. Additionally, the authors also sincerely thank all the contributors for granting permission to reprint figures, finally to the Department of Science and Technology (DST) of the Ministry of Science and Technology (MoS) and the Academy of Scientific and Innovative Research (AcSIR), New Delhi.

Conflict of interest

The authors declare no conflict of interest, financial or otherwise.

Author details

Manoshi Saikia[1,2], Samiran Bora[1,2] and Prasenjit Saikia[1,2]*

1 Coal and Energy Division, CSIR-North East Institute of Science and Technology, Jorhat, Assam, India

2 AcSIR-Academy of Scientific and Innovative Research, Ghaziabad, Uttar Pradesh, India

*Address all correspondence to: prasentu@gmail.com

References

[1] Breeze P. Power generation technologies. 3rd ed. Kidlington: Elsevier; 2019. pp. 351-3732019

[2] Mathews JP, Miller BG, Song C, Schobert HH, Botha F, Finkleman RB. The EBB and flow of US coal research 1970-2010 with a focus on academic institutions. Fuel. 2013;**105**:1-2

[3] Luttrell GH, Kohmuench JN, Yoon RH. An evaluation of coal preparation technologies for controlling trace element emissions. Fuel Processing Technology. 2000;**65**:407-422

[4] Resource C. The Coal Resource-A Comprehensive Overview of Coal, The Coal Resource-Where Does Coal Come From? What Is It. Vol. 2005. Putney: World Coal Institute; 2005. pp. 1-44

[5] Cicek T. Dry cleaning of Turkish coal. Energy Sources, Part A: Recovery, Utilization, and Environmental Effects. 2008;**30**(7):593-605

[6] International Energy Agency (IEA). Control and Minimization of Coal-Fired Power Plant Emissions. Paris: IEA; 2003

[7] Matamba T, Iglauer S, Keshavarz A. A progress insight of the formation of hydrogen rich syngas from coal gasification. Journal of the Energy Institute. 2022;**105**:81-102

[8] Fan C, Jin H. A numerical study on gasification of a single char particle in supercritical water for hydrogen production. Fuel. 2020;**268**:117399

[9] Midilli A, Kucuk H, Topal ME, Akbulut U, Dincer I. A comprehensive review on hydrogen production from coal gasification: Challenges and opportunities. International Journal of Hydrogen Energy. 2021;**46**(50):25385-25412

[10] Dai F, Zhang S, Luo Y, Wang K, Liu Y, Ji X. Recent progress on hydrogen-rich syngas production from coal gasification. PRO. 2023;**11**(6):1765

[11] Ismail TM, Shi M, Xu J, Chen X, Wang F, El-Salam MA. Assessment of coal gasification in a pressurized fixed bed gasifier using an ASPEN plus and Euler-Euler model. International Journal of Coal Science & Technology. 2020;**7**:516-535

[12] Choudhury HA, Chakma S, Moholkar VS. Biomass gasification integrated Fischer-Tropsch synthesis: Perspectives, opportunities andchallenges. In: Pandey A, Stöcker M, editors. Recent Advances in Thermochemical Conversion of Biomass. 1st ed. Netherlands: Elsevier; 2015. pp. 383-435. DOI: 10.1016/B978-0-444-63289-0.00014-4

[13] Mathekga HI, Oboirien BO, North BC. A review of oxy-fuel combustion in fluidized bed reactors. International Journal of Energy Research. 2016;**40**(7):878-902

[14] Ghassemi H, Mostafavi SM, Shahsavan-Markadeh R. Modeling of high-ash coal gasification in an entrained-flow gasifier and an IGCC plant. Journal of Energy Engineering. 2016;**142**(4):04015052

[15] Shen Z, Nikolic H, Caudill LS, Liu K. A deep insight on the coal ash-to-slag transformation behavior during the entrained flow gasification process. Fuel. 2021;**289**:119953

[16] Dorofeenko SO, Polianczyk EV. Gasification of pulverized coal in a

counterflow moving bed filtration combustion reactor: In search of the optimum. Fuel. 2021;**291**:120255

[17] Świeboda T, Krzyżyńska R, Bryszewska-Mazurek A, Mazurek W, Czapliński T, Przygoda A. Advanced approach to modeling of pulverized coal boilers for SNCR process optimization-review and recommendations. International Journal of Thermofluids. 2020;7:100051

[18] Islam MM, Hasanuzzaman M, Pandey AK, Rahim NA. Modern energy conversion technologies. In: Energy for Sustainable Development. 1st ed. Cambridge: Academic Press; 2020. pp. 19-39. DOI: 10.1016/C2017-0-01639-7

[19] Shahabuddin M, Kibria MA, Bhattacharya S. Evaluation of high-temperature pyrolysis and CO_2 gasification performance of bituminous coal in an entrained flow gasifier. Journal of the Energy Institute. 2021;**94**:294-309

[20] Chen J, Wang Q, Xu Z, Jiaqiang E, Leng E, Zhang F, et al. Process in supercritical water gasification of coal: A review of fundamentals, mechanisms, catalysts and element transformation. Energy Conversion and Management. 2021;**237**:114122

[21] Wang R, Lu L, Zhang D, Wei W, Jin H, Guo L. Effects of alkaline metals on the reactivity of the carbon structure after partial supercritical water gasification of coal. Energy & Fuels. 2020;**34**(11):13916-13923

[22] Fan C, Guo S, Jin H. Numerical study on coal gasification in supercritical water fluidized bed and exploration of complete gasification under mild temperature conditions. Chemical Engineering Science. 2019;**206**:134-145

[23] Fan C, Jin H, Shang F, Feng H, Sun J. Study on the surface structure development of porous char particles in catalytic supercritical water gasification process. Fuel Processing Technology. 2019;**193**:73-81

[24] Chen J, Wang L, Cheng Z, Lu L, Guo L, Jin H, et al. Performance simulation and thermodynamics analysis of hydrogen production based on supercritical water gasification of coal. International Journal of Hydrogen Energy. 2021;**46**(56):28474-28485

[25] Faki E, Üzden ŞT, Seçer A, Hasanoğlu A. Hydrogen production from low temperature supercritical water Co-gasification of low rank lignites with biomass. International Journal of Hydrogen Energy. 2022;**47**(12):7682-7692

[26] Ge Z, Guo L, Jin H. Catalytic supercritical water gasification mechanism of coal. International Journal of Hydrogen Energy. 2020;**45**(16):9504-9511

[27] Jin H, Fan C, Guo L, Zhao X, Cao C, Wei W. Pore structure and its evolution in char during supercritical water gasification process. Journal of Porous Media. 2019;**22**(2):195-207

[28] Feng H, Sun J, Jin H, Kou J, Guo L. Char suppression mechanism using recycled intermediate phenol in supercritical water gasification of coal. Fuel. 2021;**305**:121441

[29] Feng H, Ren Y, Fan C, Lu L, Wei W, Jin H, et al. Regulation mechanism of coal gasification in supercritical water for hydrogen production: A ReaxFF-MD simulation. International Journal of Hydrogen Energy. 2022;**47**(73):31255-31268

[30] Feng H, Kou J, Wang C, Jin H, Guo L. Insight into the internal relevance of

three-phase products in supercritical water gasification of coal. Fuel. 2022;**327**:125026

[31] Liu S, Guo L, Jin H, Li L, Li G, Yu L. Hydrogen production by supercritical water gasification of coal: A reaction kinetic model including nitrogen and sulfur elements. International Journal of Hydrogen Energy. 2020;**45**(56):31732-31744

[32] Ibrahimoglu B, Yilmazoglu MZ. Numerical modeling of a downdraft plasma coal gasifier with plasma reactions. International Journal of Hydrogen Energy. 2020;**45**(5):3532-3548

[33] Sakhraji M, Ramos A, Monteiro E, Bouziane K, Rouboa A. Plasma gasification process using computational fluid dynamics modeling. Energy Reports. 2022;**8**:1541-1549

[34] Okati A, Khani MR, Shokri B, Monteiro E, Rouboa A. Parametric studies over a plasma co-gasification process of biomass and coal through a restricted model in Aspen plus. Fuel. 2023;**331**:125952

[35] Pan P, Peng W, Li J, Chen H, Xu G, Liu T. Design and evaluation of a conceptual waste-to-energy approach integrating plasma waste gasification with coal-fired power generation. Energy. 2022;**238**:121947

[36] Messerle VE, Ustimenko AB, Lavrichshev OA. Comparative study of coal plasma gasification: Simulation and experiment. Fuel. 2016;**164**:172-179

[37] Popov VE, Subbotin DI, Surov AV, Popov SD, Serba EO, Godina EP, et al. Co-gasification of lignite and used car tires by H_2O/air thermal plasma. Journal of Physics: Conference Series. 2019;**1243**(1):012010

[38] Wei G, Zhao W, Huang Z, Yi Q, He F, Zhao K, et al. Synthesis gas production from chemical looping gasification of lignite by using hematite as oxygen carrier. Energy Conversion and Management. 2019;**185**:774-782

[39] Zhao K, Fang X, Huang Z, Wei G, Zheng A, Zhao Z. Hydrogen-rich syngas production from chemical looping gasification of lignite by using $NiFe_2O_4$ and $CuFe_2O_4$ as oxygen carriers. Fuel. 2021;**303**:121269

[40] Meng F, Li X, Qiu S, Li J, Wang H, Hu G. Study on hydrogen production of ZhunDong coal using chemical looping with Cu-Fe as oxygen carrier. Combustion Science and Technology. 2022;**194**(15):3188-3205

[41] Cui Z, Sun S, Zhang H, Liu B, Tian W, Guo Q. Comprehensive optimization of coal chemical looping gasification process for low CO_2 emission based on multi-scale simulation coupled experiment. Fuel. 2022;**324**:124464

[42] Cui Z, Tian W, Zhang H, Guo Q. Multi-scale modeling and control of chemical looping gasification coupled coal pyrolysis system for cleaner production of synthesis gas. Journal of Cleaner Production. 2021;**299**:126903

[43] Xie F, An M, Li P, Hu X, Bai H, Guo Q. Simulation study on the gasification process of Ningdong coal with iron-based oxygen carrier. Chinese Journal of Chemical Engineering. 2021;**29**:326-334

[44] Zhu X, Zhang J, Yan J, Shen L. Characteristic evaluation and process simulation of CuFe2O4 as oxygen carriers in coal chemical looping gasification. ACS Omega. 2021;**6**(7):4783-4792

[45] An M, Guo Q, Ma J, Hu X. Chemical-looping gasification of coal with

$CuFe_2O_4$ oxygen carriers: The reaction characteristics and structural evolution. The Canadian Journal of Chemical Engineering. 2020;**98**(7):1512-1524

[46] Kun Z, He D, Guan J, Shan L, Wu Z, Zhang Q. Coal gasification using chemical looping with varied metal oxides as oxygen carriers. International Journal of Hydrogen Energy. 2020;**45**(18):10696-10708

[47] Wei G, Yang M, Huang Z, Bai H, Chang G, He F, et al. Syngas production from lignite via chemical looping gasification with hematite oxygen carrier enhanced by exogenous metals. Fuel. 2022;**321**:124119

[48] Kun Z, He D, Guan J, Wang Q, Li X, Shang J, et al. Interaction between bimetallic composite oxygen carriers and coal and its contribution to coal direct chemical looping gasification. International Journal of Hydrogen Energy. 2020;**45**(38):19052-19066

[49] Yang J, Ren Y, Lu J, Liu H, Zhang Z, Pang H, et al. Chemical looping gasification with a $CuFe_2O_4$-enhanced phosphogypsum oxygen carrier during reduction in a fluidized bed reactor. Chemical Engineering Journal. 2021;**426**:131346

[50] Meng F, Li X, Qiu S, Li J, Wang H, Hu G. Study on hydrogen production of Zhun Dong coal using chemical looping with Cu-Fe as oxygen carrier. Combustion Science and Technology. 2022;**194**(15):3188-3205

[51] Zhang J, Wang Y, Dong L, Gao S, Xu G. Decoupling gasification: Approach principle and technology justification. Energy & Fuels. 2010;**24**(12):6223-6232

[52] Liu Q, Zhao YJ, Huang Y, Pei F, Cui Y, Shi LJ, et al. Pilot test of low-rank coal pyrolysis coupled with gasification to hydrogen-rich gas for direct reduced iron: Process modeling, simulation and thermodynamic analysis. Fuel. 2023;**331**:125862

[53] Wang C, Zhang M, Han Z, Bai D, Duo W, Bi X, et al. Pilot verification of a two-stage fluidized bed gasifier with a downer pyrolyzer using oxygen-rich air. Fuel. 2022;**307**:121816

[54] He S, Li S, Gao L. Proposal and energy saving analysis of novel methanol–electricity polygeneration system based on staged coal gasification method. Energy Conversion and Management. 2021;**233**:113931

[55] He S, Gao L, Dong R, Li S. A novel hydrogen production system based on the three-step coal gasification technology thermally coupled with the chemical looping combustion process. International Journal of Hydrogen Energy. 2022;**47**(11):7100-7112

[56] Zhang B, Yang B, Guo W, Wu S, Zhang J, Wu Z. Chemical looping gasification of maceral from low-rank coal: Products distribution and kinetic analysis on vitrinite. Chinese Journal of Chemical Engineering. 2021;**36**:233-241

[57] Pelissari MR, Cañas SS, Barbosa MO, Tassinari CC. Decarbonizing coal-fired power plants: Carbon capture and storage applied to a thermoelectric complex in Brazil. Results in Engineering. 2023;**19**:101249. Available from: https://www.iea.org/energy-system/carbon-capture-utilisation-and-storage/co2-capture-and-utilisation

[58] Butalia T, Wolfe W, Amaya P. The utilization of flue-gas desulfurization materials. In: Coal Combustion Products (CCP's). 1st ed. New Delhi. Woodhead Publishing; 2017. pp. 155-184

[59] Córdoba P. Status of flue gas desulphurisation (FGD) systems from coal-fired power plants: Overview of the physic-chemical control processes of wet limestone FGDs. Fuel. 2015;**144**:274-286

[60] Su Q, Robl TL, Jewell RB, Wolfe WE, Butalia TS, Zhao Z, et al. Development of key technologies for dry FGD by-product resource comprehensive utilization: A project history. In: 2015 World of Coal Ash (WOCA) Conference, Nashville, TN, May 5-7. 2015

[61] Singh DK, Tirkey JV. Performance optimization through response surface methodology of an integrated coal gasification and CI engine fuelled with diesel and low-grade coal-based producer gas. Energy. 2022;**238**:121982

[62] Gupta PK, Kumar V, Maity S. Renewable fuels from different carbonaceous feedstocks: A sustainable route through Fischer–Tropsch synthesis. Journal of Chemical Technology & Biotechnology. 2021;**96**(4):853-868

[63] Martin O, Martín AJ, Mondelli C, Mitchell S, Segawa TF, Hauert R, et al. Indium oxide as a superior catalyst for methanol synthesis by CO_2 hydrogenation. Angewandte Chemie. 2016;**128**(21):6369-6373

[64] Wang D, Li S, He S, Gao L. Coal to substitute natural gas based on combined coal-steam gasification and one-step methanation. Applied Energy. 2019;**240**:851-859

[65] Ye C, Ye Z, Zhu Z, Wang Q. Thermodynamic and economic analysis of oxy-fuel-integrated coal partial gasification combined cycle. ACS Omega. 2021;**6**(6):4262-4272

[66] Diao R, Li S, Deng J, Zhu X. Interaction and kinetic analysis of co-gasification of bituminous coal with walnut shell under CO_2 atmosphere: Effect of inorganics and carbon structures. Renewable Energy. 2021;**173**:177-187

[67] Yang X, Liang Z, Chen H, Wang J, Mu X. Effects of K_2CO_3 and Ca (OH)2 on CO_2 gasification of char with high alkali and alkaline earth metal content and study of different kinetic models. Thermal Science. 2022;**26**(1 Part A):119-133

[68] Lv P, Wang J, Bai Y, Song X, Su W, Yu G, et al. CO_2 gasification of petroleum coke with use of iron-based waste catalyst from FT synthesis. Thermochimica Acta. 2022;**711**:179205

[69] Ellison C, Abdelsayed V, Smith M, Shekhawat D. Comparative evaluation of microwave and conventional gasification of different coal types: Experimental reaction studies. Fuel. 2022;**321**:124055

[70] Lu G, Bai Y, Lv P, Wang J, Song X, Su W, et al. Changes in active sites and reactivity induced by interactions among Ca, Si and Al during coal char gasification. Fuel. 2022;**310**:122322

[71] Hong D, Wang Y, Zhai X. A ReaxFF study of the effect of pressure on the contribution of char-CO_2 gasification to char conversion during pressurized oxy-fuel combustion. Fuel. 2022;**329**:125439

[72] Qiu Q, Pan D, Zhang W, Zeng F, Liu L. Catalytic kinetics and mechanisms of KCl with different concentrations on gasification of coal char. PRO. 2022;**10**(7):1357

[73] Lu Q, Guo R, Zhang H, Wang JP, Lu T, Li GY, et al. To stimulate, and to inhibit: A theoretical understanding of the sodium-catalytic mechanism of coke gasification. Chemical Engineering Journal. 2022;**435**:135091

[74] Li L, Wang Z, Zhao R, Mei Y, Shi W, Liu Z, et al. The different catalytic effects of Na species on char gasification and the reasons for this different. Journal of Thermal Analysis and Calorimetry. 2022;**147**(10):5687-5699

[75] Yu G, Yu D, Liu F, Han J, Yu X, Wu J, et al. Different impacts of magnesium on the catalytic activity of exchangeable calcium in coal gasification with CO_2 and steam. Fuel. 2020;**266**:117050

[76] Yao Q, Ma M, Liu Y, He L, Sun M, Ma X. Pyrolysis characteristics of metal ion-exchanged Shendong coal and its char gasification performance. Journal of Analytical and Applied Pyrolysis. 2021;**155**:105087

[77] Mao L, Li H, Xia B, Liu T, Zhang Y, Zheng M. Effect of Ca/Fe-based auxiliaries on anthracite char gasification under CO_2 atmosphere: Synergistic catalysis. Thermochimica Acta. 2022;**713**:179224

[78] Malik A, Chaturvedi V, Sandhani M, Das P, Arora C, Singh N, et al. Implications of an emission trading scheme for India's net-zero strategy: A modelling-based assessment. Environmental Research Letters. 30 Jul 2024;**19**(8):084043

[79] Sasol Performance Chemicals. Historical landmarks. [Internet] 2020. Available from: https://www.sasol.com/about-sasol/company-profile/historical-milestones

[80] Höök M, Aleklett K. A review on coal-to-liquid fuels and its coal consumption. International Journal of Energy Research. 2010;**34**(10):848-864

[81] Nalwala RG, Rana PK, Shukla DH. Kinetic Modelling for Coal Liquefaction Process. International Journal of Engineering Research & Technology (IJERT). 2014;**3**(5)

[82] World Coal Institute – UNECE, International association of coal enterprises & stakeholders, Geneva. 2-3 Feb 2006

[83] Ali A, Zhao C. Direct liquefaction techniques on lignite coal: A review. Chinese Journal of Catalysis. 2020;**41**(3):375-389. ISSN 1872-2067. DOI: 10.1016/S1872-2067(19)63492-3

[84] Gonçalves RV, Vono LL, Wojcieszak R, Dias CS, Wender H, Teixeira-Neto E, et al. Selective hydrogenation of CO_2 into CO on a highly dispersed nickel catalyst obtained by magnetron sputtering deposition: A step towards liquid fuels. Applied Catalysis B: Environmental. 2017;**209**:240-246

[85] A Barrel Full. Veba Combi Cracking VCC [Internet] 2013. Available from: http://abarrelfull.wikidot.com/veba-combi-cracking-vcc [Accessed: May 01, 2020]

[86] Miller BG. Introduction to coal utilization technologies. In: Clean Coal Engineering Technology. 1st ed. Oxford: Butterworth-Heinemann; 2017. pp. 147-229

[87] Daranin EA, Umar DF, Zulfahmi Z, Wijaya T, Setiawan L. Preliminary study of the methanol production process from coal to support the coal downstream program. In: IOP Conference Series: Earth and Environmental Science. IOP Publishing. 2023;**1288**(1):012006

[88] Wang Q, Wang E, Li K, Husnain N, Li D. Synergistic effects and kinetics analysis of biochar with semi-coke during CO_2 co-gasification. Energy. 2020;**191**:116528

[89] Beck A, Newton MA, van de Water LG, van Bokhoven JA. The enigma of methanol synthesis by Cu/ZnO/Al2O_3-based catalysts. Chemical Reviews. 2024;**124**(8):4543-4678

[90] National Energy Technology Laboratory. Gasification [Internet]. 2020. Available from: https://www.netl.doe.gov/research/Coal/energysystems/gasification/gasifipedia

[91] Xu X, Liu Y, Zhang F, Di W, Zhang Y. Clean coal technologies in China based on methanol platform. Catalysis Today. 2017;**298**:61-68

[92] Li R, Yang Z, Duan Y. Energy, economic and environmental performance evaluation of co-gasification of coal and biomass negative-carbon emission system. Applied Thermal Engineering. 2023;**231**:120917

[93] Shahabuddin M, Bhattacharya S. Enhancement of performance and emission characteristics by co-gasification of biomass and coal using an entrained flow gasifier. Journal of the Energy Institute. 2021;**95**:166-178

[94] Jiao Z, Chen X, Zhao Y, Liu L, Xing C, Zhang L, et al. Effect of active alkali and alkaline earth metals on the reactivity of co-gasification char from coal and corn straws. Journal of the Energy Institute. 2022;**102**:42-53

[95] Barontini F, Frigo S, Gabbrielli R, Sica P. Co-gasification of woody biomass with organic and waste matrices in a down-draft gasifier: An experimental and modeling approach. Energy Conversion and Management. 2021;**245**:114566

[96] Dhrioua M, Ghachem K, Hassen W, Ghazy A, Kolsi L, Borjini MN. Simulation of biomass air gasification in a bubbling fluidized bed using aspen plus: A comprehensive model including tar production. ACS Omega. 2022;**7**(37):33518-33529

[97] Dhrioua M, Hassen W, Kolsi L, Ghachem K, Maatki C, Borjini MN. Simulation of Prosopis juliflora air gasification in multistage fluidized process. PRO. 2020;**8**(12):1655

[98] Ashraf A, Sattar H, Munir S. A comparative performance evaluation of co-combustion of coal and biomass in drop tube furnace. Journal of the Energy Institute. 2022;**100**:55-65

[99] Gray RJ. Coal to coke conversion. In: Introduction to Carbon Science. Oxford: Butterworth-Heinemann; 1989. p. 285-321

[100] Bedenko SP, Dement'ev KI, Maximov AL. Modern processes for petrochemistry based on acetylene (A review). Petroleum Chemistry. 2022;**62**(9):989-1026

[101] Huang Y, Feng J, Liang CH, Huang P, Zhang XW, Xie Q, et al. Co-production of naphthenic oil and phenolic compounds from medium-and low-temperature coal tar. Industrial & Engineering Chemistry Research. 2021;**60**(16):5890-5902

[102] Zhang Y, Li H, Ramirez Reina T, Liu J. Coal chemistry industry: From production of liquid fuels to fine chemicals to carbon materials. Energy & Fuels. 2023;**37**(23):17754-17764

[103] Lee WH, Lee CW, Cha GD, Lee BH, Jeong JH, Park H, et al. Floatable photocatalytic hydrogel nanocomposites for large-scale solar hydrogen production. Nature Nanotechnology. 2023;**18**(7):754-762

[104] Ye M, Tian P, Liu Z. DMTO: A sustainable methanol-to-olefins technology. Engineering. 2021;7(1):17-21

[105] Zhang Y, Zhang K, Jia K, Liu G, Ren S, Li K, et al. Preparation of coal-based graphene quantum dots/α-Fe_2O_3 nanocomposites and their lithium-ion storage properties. Fuel. 2019;**241**:646-652

[106] Lv X, Zhang T, Luo Y, Zhang Y, Wang Y, Zhang G. Study on carbon nanotubes and activated carbon hybrids by pyrolysis of coal. Journal of Analytical and Applied Pyrolysis. 2020;**146**:104717

[107] Luo Z, Ma J, Wang X, Chen D, Wu D, Pan J, et al. Surface engineering of fluorinated graphene nanosheets enables ultrafast lithium/sodium/potassium primary batteries. Advanced Materials. Oct 2023;**35**(40):2303444

[108] Wang L, Sun F, Gao J, Zhu Y, Pei T, Li L, et al. Pore reorganization of porous carbon during trace calcium-catalyzed coal activation for adsorption applications. Energy & Fuels. 2018;**32**(9):9191-9201

[109] Xiao N, Wei Y, Li H, Wang Y, Bai J, Qiu J. Boosting the sodium storage performance of coal-based carbon materials through structure modification by solvent extraction. Carbon. 2020;**162**:431-437

[110] International Energy Agency. Energy Supply and Consumption Revolution Strategy (2016-2030). International Energy Agency; 16 April 2021. Available from: https://www.iea.org/policies/1794-energy-supply-and-consumption-revolution-strategy-2016-2030

[111] European Consensus on Development. Available from: https://ec.europa.eu/europeaid/policies/european-development-policy/2030-agenda-sustainabledevelopment_en

[112] Focus on 100 million Ton Coal Gasification by 2030 [Internet] 2023. Information Bureau Government of India. Date-26 JUL 2023. Available from: https://pib.gov.in/PressReleasePage.aspx?PRID=1942811]

[113] Global coal to liquid market by technology (direct liquefaction, indirect liquefaction), by product(diesel, gasoline, jet fuel, chemicals, others), by application(transportation fuel, chemicals and petrochemicals, power generation, cooking fuel, others), by region and companies - industry segment outlook, market assessment, competition scenario, trends, and forecast 2024-2033 [Internet]. 2024. Available from: https://www.sphericalinsights.com/reports/coal-to-liquid-market

[114] Carbon capture and storage market size, share & trends analysis report by technology (pre-combustion capture, oxy-fuel combustion capture, post-combustion capture), by end-user industry (oil and gas, coal and biomass power plant, iron and steel, chemical) and by region (North America, Europe, APAC, Middle East and Africa, LATAM) Forecasts, 2023-2031. 2024. Available from: https://straitsresearch.com/report/carbon-capture-and-storage-market#:~:text=Market%20Overview,period%20(2022%2D2031

[115] National Energy Technology Laboratory. Summary Information for External R&D Awards. 2024. Available from: https://www.netl.doe.gov/node/2476

[116] Muller-Feuga A. The Recognition of Carbon Capture and Storage by Plants. Journal of Agricultural Science. 2024;**16**(7)

[117] Livescu A, Navar R, Mangalindan JR, Mahnaz F, Ge Y, Shetty M, et al. Catalysts for clean energy: A review on current progress for the catalyzed recycling of CO_2 into dimethyl ether. Topics in Catalysis. 2024:1-22

Chapter 6

Asphaltite Potential Assessment as Energy Source in Şırnak

Edip Taşkesen and Hamza Alahmad

Abstract

Asphaltites are solid hydrocarbons, typically black, with a matte or shiny with an average calorific value of 4500 kcal/kg. Asphaltites are suitable for electricity generation; however, due to their high ash and sulfur content, they are not ideal for domestic use. Minerals and by-products, such as synthetic oil, ammonia, and H_2S can be obtained from asphaltites. This mineral is a rare yet new energy source, especially for Turkey, where it has significant reserves available (82 million tons). The thermal power plant in the Silopi district of Turkey serves as a new technology trial for asphalted-based energy production (Fluidized-Bed Combustion or FBC technology). This chapter provides important data for future potential energy assessment opportunities and has innovative content for the industry.

Keywords: asphaltite, mining, energy, thermal power plant, fluidized bed combustion FBC

1. Introduction

The rapid increase in population and industrialization has significantly escalated energy demand, particularly in developing countries [1, 2]. Economic and social development are among the primary indicators of a nation's growth and welfare. As social development and economic prosperity improve, energy consumption tends to rise proportionally [3]. As of 2022, global energy consumption reached 183,230 TWh, with 81.4% sourced from fossil fuels, highlighting the world's dependency on nonrenewable energy resources (**Figure 1**).

According to the 2022 report by the Turkish Statistical Institute (TÜİK), Turkey's population growth rate decreased from 12.7‰ in 2021 to 7.1‰ in 2022. This decline reflects a slowdown in the country's population expansion, yet Turkey continues to rank among the most populous nations in Europe, highlighting its significant demographic presence on the continent [6, 7]. The country fulfills a substantial portion of its energy needs through fossil fuels. In 2023, Turkey's total energy consumption was 1945 KWh (167.3 Mtoe), and electricity consumption reached 323.8 TWh, marking a 1.5% increase from the previous year. Globally, coal accounts for 35.5% of energy production, whereas in Turkey, this share rises to 37% (**Figure 2**) [4, 8]. Given Turkey's existing coal reserves, this percentage is likely to increase. The utilization of coal in Turkey is divided as follows: 76% in thermal power plants, 14% in industry, and 10% for residential heating purposes [9].

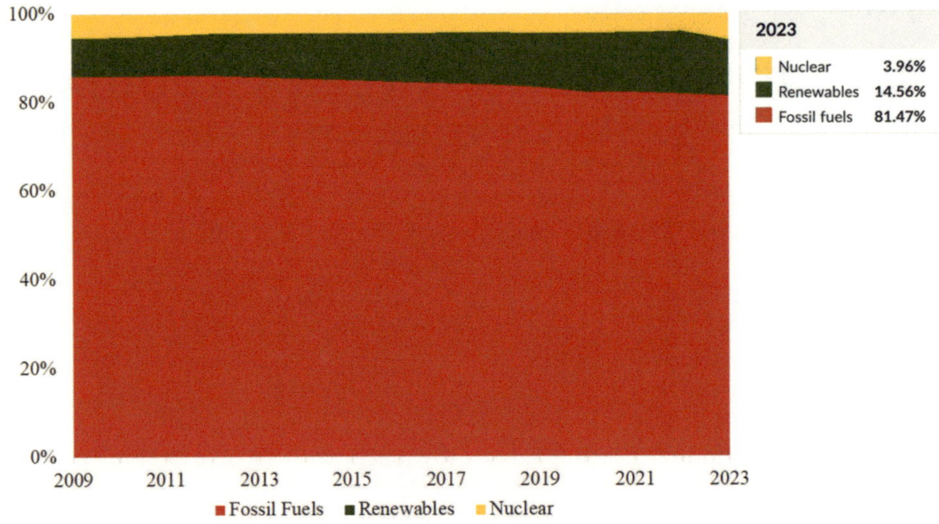

Figure 1.
Share of world primary energy consumption by source type from 2009 to 2023 [4, 5].

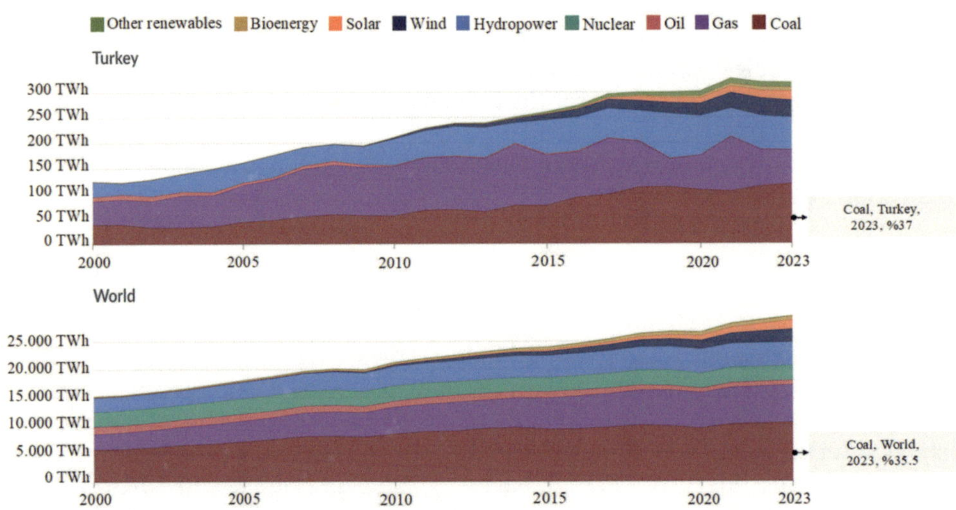

Figure 2.
Proportion of energy sources used in electricity generation in Turkey and the world between 2000 and 2023 [8].

Energy shortages remain a critical issue globally and in Turkey. Reducing energy dependency through the effective utilization of domestic energy resources is essential. Research has revealed that asphaltite can also be employed as an industrial raw material. The combustion of asphaltite produces ash that contains valuable elements such as nickel, molybdenum, vanadium, and uranium. In Turkey's Eastern and Southeastern Anatolia Regions, asphaltite has found applications in heating and, more recently, in electricity generation through thermal power plants. The Şırnak-Silopi thermal power plant, which became fully operational in 2015, produces 405 MW of electricity [10].

Turkey's most significant asphaltite reserves are in two main fields in the Southeastern Anatolia Region: one south of Şırnak and the other northeast of Silopi.

Drilling evaluations conducted by the General Directorate of Mineral Research and Exploration (MTA) estimate Turkey's total asphaltite reserves to be approximately 82 million tons. Traditionally, asphaltite produced in Turkey has been used for residential heating in Eastern and Southeastern Anatolia. However, due to its high sulfur content, its use in residential areas poses environmental and health risks [11]. As a result, studies have focused on improving the quality of asphaltite and reducing its sulfur content through purification processes.

This chapter evaluates the formation, reserves, applications, and purification processes of asphaltite, offering insights into its potential as an energy resource.

2. Definition, types, and chemical properties of asphaltenes

Hydrocarbons with origins similar to crude oil are naturally found underground, often formed during the metamorphism of petroleum due to environmental factors or during the formation of oil reservoirs. Two significant formations resulting from these processes are asphalt lakes and bituminous sands [3, 11]. These hydrocarbons primarily occur in two structural forms: asphaltite and asphaltic pyrobitumen, which occupy rock fractures or vein systems. These materials undergo hydrogen depletion and carbon enrichment, leading to the formation of natural asphalt, asphaltite, and asphaltic pyrobitumen [3]. **Figure 3** shows the asphaltic substances formed by the metamorphism of oil. **Figure 4** shows the general appearance of asphaltenes.

Asphaltenes are organic substances found in nature, typically black and shiny in appearance, with a variety of types based on their carbon content [11, 13]. The main types of asphaltenes include Jilsonite, Glance Pitch, and Grahamite, each with distinct chemical compositions and physical properties. These types of asphaltenes

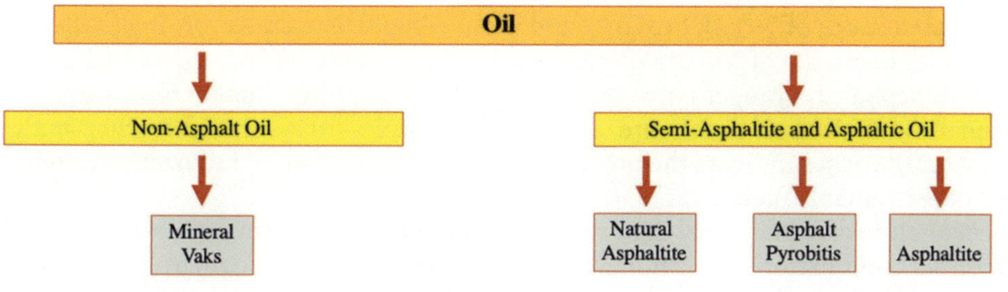

Figure 3.
Asphaltic substances resulting from metamorphism from oil [12].

Figure 4.
Asphaltite appearance [13].

Asphaltite type	Fixed carbon ratio (%)
Jilsonite	10–20
Glance pitch	20–30
Grahamit	30–55

Table 1.
Types of asphaltites according to fixed carbon percentages [12, 14].

vary in their carbon content and can be found as natural resources in regions such as the United States and Turkey. The carbon content of these asphaltenes varies significantly (**Table 1**) [15].

Asphaltites are solid hydrocarbons, typically black with a matte or shiny appearance. Their hardness measures 2–3 on the Mohs scale, and their specific gravity ranges from 1.03 to 1.20. The calorific value of asphaltites lies between 11 and 24 MJ/kg (2388–5493 kcal/kg) [14]. Asphaltites demonstrate high solubility in carbon disulfide (CS_2), with solubility levels reaching 90–100%, and their melting points range between 120 and 315°C. Chemically, they primarily consist of carbon and hydrogen, while the mineral content may constitute up to 40% of their composition [16].

Deposits from the Mesozoic-Senozoic geological periods often include clay, quartz, albite, orthoclase, and pyrite minerals [17, 18]. These formations are commonly found in structural fractures associated with fault displacements during the Tertiary period [19]. Asphaltites are characterized by high ash content, which may contain trace metals such as molybdenum (Mo), nickel (Ni), and titanium (Ti) [20, 21]. Compared to coal types like lignite, asphaltites have a notably lower oxygen content (approximately 3%), whereas oxygen content in lignite ranges from 3–44% [11].

In Turkey, economically significant asphaltite and asphaltic pyrobitumen reserves are located south of Şırnak Province and in the Silopi District. According to the General Directorate of Mineral Research and Exploration (MTA), the calorific value of these asphaltites ranges between 3000 and 6000 kcal/kg. Their composition typically includes 1–2% moisture, 4–7% sulfur, and 35–40% ash. Laboratory analyses of asphaltite ashes indicate the presence of rare earth elements, radioactive minerals, and other valuable metals [11, 13].

Asphaltites are primarily used as a source of energy in thermal power plants and for direct combustion. Their mineralogical properties also make them suitable for industrial and metallurgical applications, highlighting their potential for broader utilization [22].

2.1 Jilsonite asphaltene

Jilsonite is a shiny black asphaltene with a conchoidal fracture, resembling obsidian. A typical example is found in the Uinta Basin of Utah, where veins crosscut the Uinta Formation. The solubility of Jilsonite in carbon disulfide (CS2) ranges from 95 to 98%, and its melting temperature ranges from 161 to 230°C. Its density is approximately 1.03–1.10 g/cm^3 [23]. A brief analysis of its physical properties is given in **Table 2**, and its chemical composition is shown in **Table 3. Figures 4–6** depict the physical appearance of Jilsonite.

Jilsonite is commonly used in the construction industry, particularly as an insulation material in paint and ink production. This valuable resource is not only found in the United States but also in countries such as Canada, Mexico, Iran, and China,

Jilsonite asphaltite physical properties						
C (%)	**Ash (%)**	**Moisture (%)**	**Volatile substance (%)**	**Softening point (°C)**	**Melting degree (°C)**	**Solubility in H_2S (%)**
43	2,4	0,9	53,7	132–190	161–230	95–98

Table 2.
Physical characteristics of jilsonit [24].

Jilsonit asphaltite chemical properties (%)				
C	**H**	**N**	**S**	**O**
81	6,7	0,84	2,1	2,3

Table 3.
Chemical composition of jilsonite [24].

Figure 5.
Physical appearance of jilsonite [25].

Figure 6.
Jilsonite asphaltite [26].

which have natural Jilsonite deposits. These deposits play a significant role in providing the raw material for industrial applications, and Jilsonite mining is an important economic activity in these regions. Specifically, Jilsonite is used as a binder and additive in paint and ink production to improve durability and color quality, while in the insulation industry, it is added to building materials to enhance energy efficiency [3]. Thus, Jilsonite is a versatile and valuable natural resource with a significant role in global economies (**Figure 6**).

2.2 Glance pitch asphaltene

Glance pitch is another shiny black asphaltene with a conchoidal fracture. This type has a density of 1.10–1.15 g/cm^3, a fixed carbon content of 20–30%, and a sulfur content ranging from 2 to 8%. Its ash content is low, and its melting temperature ranges from 110 to 121°C. Like Jilsonite, glance pitch asphaltene is highly soluble in carbon disulfide, ranging from 97 to 100%. This asphaltene is known as "manjak" and can be found in countries like Cuba, Mexico, Nicaragua, El Salvador, Colombia, Argentina, Chile, Syria, and Iraq [11, 22]. The chemical composition of glance pitch asphaltene is shown in **Table 4**.

2.3 Grahamite asphaltene

Grahamite differs from other asphaltene types in some key properties. Its density ranges from 1.15 to 1.50 g/cm^3, and its fixed carbon content is between 30 and 55%. The melting temperature ranges from 175 to 315°C, and its solubility in carbon disulfide is between 90 and 100%. Grahamite asphaltene is found in regions such as West Virginia, Oklahoma, Mexico, Canada, Cuba, Argentina, and Peru, and in Turkey, asphaltites are predominantly of the grahamite type [3, 23]. **Table 5** summarizes the characteristics of the various Grahamit.

2.4 Applications and importance of asphaltenes

Asphaltenes are highly valuable in the energy industry and other sectors due to their diverse properties and benefits. Below are some of the main uses of asphaltene:

- Asphalt production: Asphaltite is an important raw material of asphalt production [3]. It is preferred in the production of high-quality asphalt used in highways, airports, and other road construction projects. This represents an important area of use in infrastructure development and maintenance projects.

Chemical properties of glance pitch (%)				
C	H	N	S	O
82	7,4	1,02	8,69	0,9

Table 4.
Chemical composition of glance pitch [24].

Asphalt properties			
Asphaltite type	Intensity $\left(g/cm^3\right)$	Melting degree (°C)	Softening point (°C)
Grahamit	1,15–1,50	175–315	188–329

Table 5.
Characteristics of grahamit [27].

- Water isolation: Asphaltite is used in the production of waterproof material [3]. It is widely preferred especially in the waterproofing of roofing, water tanks, dams, and tunnels. This plays an important role in protecting water resources and preventing water leaks.

- Cement production: Some types of asphaltites are used in cement production [28]. It is especially important as additives used in the production of cement types used in construction projects that require rapid attachment.

- Chemical industry: Asphaltite is used as a catalyst or hammade in the production of some chemical products [28]. This is necessary for many processes in the petrochemical and chemical industry.

- Energy production: Asphaltite can be used in thermal energy production [29]. It can serve as an alternative in energy production, especially in plants with high energy needs, using it as fuel.

- These various uses of asphaltite make it an essential component of the energy sources and construction industries. Beyond that, the importance of asphaltite can be explained by:

- Infrastructure development: Contributes to the successful realization of infrastructure development projects such as asphaltite, road construction, waterproofing and construction projects [3]. These projects are critical to economic development and social welfare.

- Sustainability: Offers sustainable solutions in areas such as asphaltite, waterproofing and energy production [3]. It can be used to protect water resources, increase energy efficiency and reduce environmental impacts.

- Economic benefit: These broad areas of use of asphaltite contribute to the creation of business opportunities and economic growth [3]. It can also reduce energy imports using local energy sources, which strengthens the economy.

Asphaltite plays an important role in various industries when used as an energy source. This mineral helps to successfully complete infrastructure projects, increase sustainability and provide economic benefits. Therefore, the potential assessment of asphaltite has a strategic importance in the energy sources and construction industries.

3. Investigation of asphaltite distribution and potential in Şırnak province

Asphaltite is a mineral consisting of hydrocarbons derived from petroleum and containing two different types, asphaltene and asphaltic pyrobitumen, which either fill rock cracks or are found in veins. These hydrocarbons become enriched in carbon as they undergo hydrogen loss, ultimately forming soft and hard components such as natural asphalt, asphaltite, and asphaltic pyrobitumen. While asphaltite and asphaltic pyrobitumens, which are widely found throughout Turkey, are economically significant in energy production, the formation of these minerals is only observed in the south of Şırnak city center and Silopi District [22]. Şırnak Province is home to one of these important asphaltite fields, and the distribution of the field is seen in various veins. Areas in the southern region of Şırnak, such as Avgamasya, Segürek, Seridahlı, Nivekara, Milli, İspindoruk, Karatepe, and Rutkekurat, are known for their asphaltite deposits, while Üçkardeşler, Harbul, and Silip fields in the northeast of Silopi also constitute significant asphaltite areas in the region [3, 30]. Additionally, the Ortasu asphaltite field, located in the southeast of Şırnak Province, near the Iraqi border, is noteworthy. According to data from the General Directorate of Mineral Research and Exploration (MTA), the total asphaltite reserve in this region is estimated to be approximately 82 million tons [29, 31]. The largest reserves are found in Harbul (25.8 million tons), Üçkardeşler (20.4 million tons), and Avgamasya (14 million tons) [32]. Asphaltites in the Şırnak region are used primarily for detection and electricity generation, especially in the Eastern and Southeastern Anatolia regions. Asphaltite studies in this area began in 1964 and continued in various periods until 1988 with geological surveys and drilling activities aimed at reserve detection, initiated by the Turkish Coal Enterprises (TKİ). A total of 12 asphaltite deposits have been identified in the Şırnak region. These phyla are located on the map shown in **Figures 7** and **8**. **Figure 9** presents a graphical representation of asphaltite reserves in Şırnak Province. The main characteristics of these deposits are summarized in **Table 6** for each deposit.

3.1 Avgamasya fleet

The Avgamasya fleet is one of the largest asphaltite veins in the Southeastern Anatolia Region, approximately 3 kilometers long and 75 meters wide. This vein cuts through the Germav formation, which dates back to the Late Cretaceous and Paleocene periods. The southwestern part of the fleet developed as a single fracture fill, while the northeastern part split and formed a series of component fleets [3].

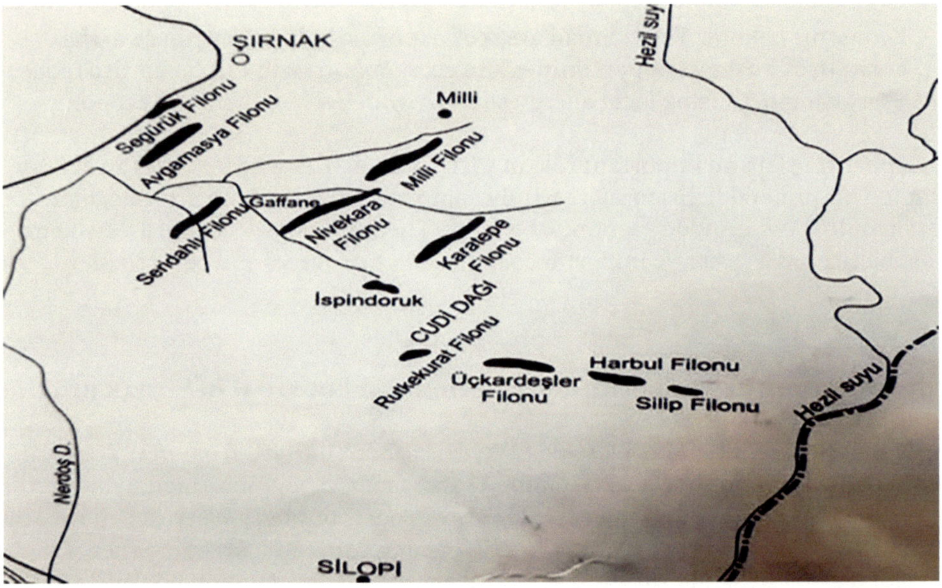

Figure 7.
Asphaltite reserves in the southeastern Anatolia region [33].

DOI: http://dx.doi.org/10.5772/intechopen.1008725

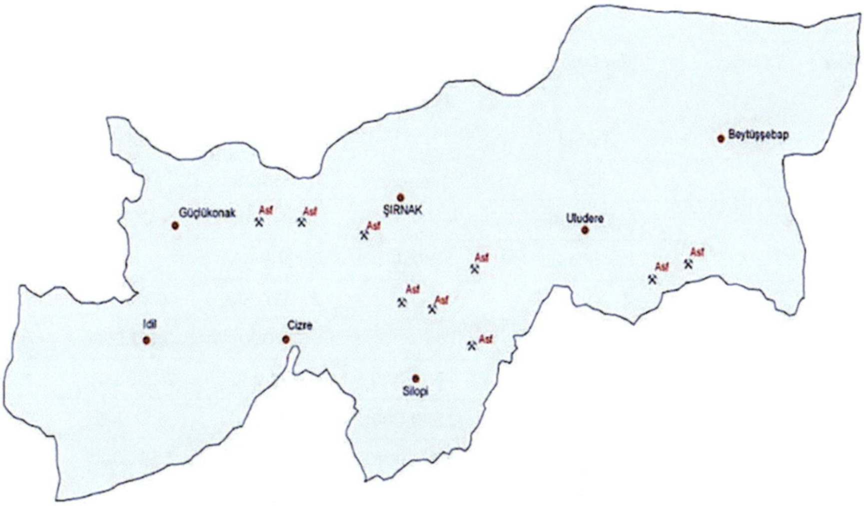

Figure 8.
Geographical locations of asphaltite resources in Şırnak province.

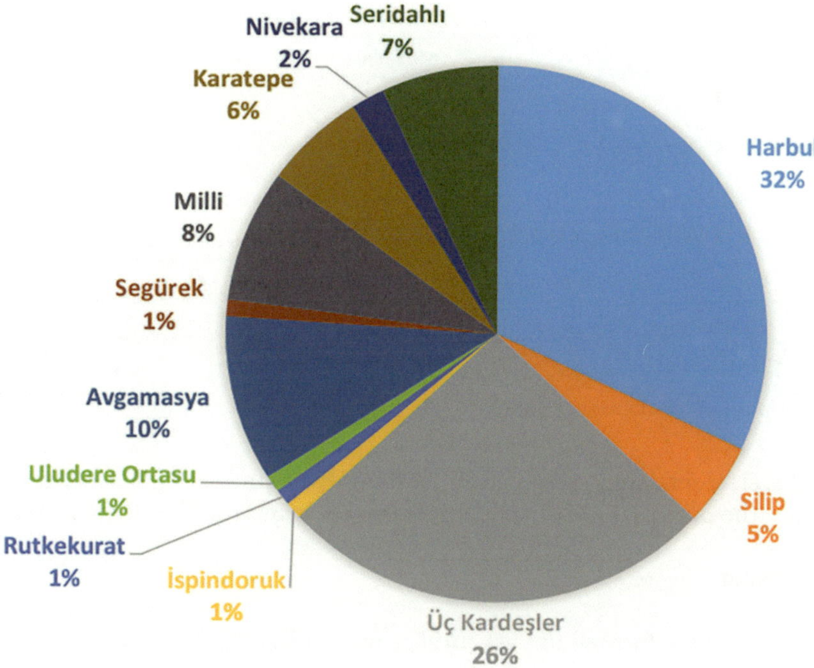

Figure 9.
Distribution rates of asphaltite reserves in Şırnak to different fleets [3, 34].

Avgamasya, with a reserve of 14 million tons, is one of Turkey's significant asphaltite fields [16, 21]. Its mineral composition includes calcite, dolomite, pyrite, quartz, and small amounts of feldspar, illite, and gypsum [36]. The average pore size of these asphaltites is 15 microns, with a pyrite content of 12%. Additionally, the Rmax value is 0.54%, and the clay and inorganic matter content is approximately 20% [37]. **Table 7**

Şırnak asphalt fields resource status						
Province	District	Fleet name	Asphaltite source (Ton)			
			Possible	Likely	Visible	Total
Şırnak	Sillopi	Harbul	—	7.851.000	17.914.000	25.765.000
		Silip	—	1.335.000	3.071.000	4.406.000
		Üçkardeşler	—	10.881.000	9.472.000	20.352.000
		Rutkekurat	1000.000	—	—	1.000.000
		Total	—	20.067.000	30.457.000	51.523.000
	Şırnak	Avgamasya	—	673.000	7.481.000	8.154.000
		Mili	1.600.000	2.900.000	2.000.000	6.500.000
		Anılmış-Karatepe	2.500.000	2.000.000	500.000	5.000.000
		Seridahili	1.279.000	1.254.000	3.534.000	6.067.000
		Nivekara	700.000	1.000.000	300.000	2.000.000
		Segürük	—	450.000	550.000	1.000.000
		Anılmış-İspindoruk	500.000	500.000	100.000	1.100.000
		Uludere Ortasu	—	53.000	551.000	604.000
		Total	6.579.000	8.830.000	15.016.000	30.425.000
Total (General)			6.579.000	28.897.000	45.473.000	81.949.000

Table 6.
Reserve information of asphaltite fleets in Şırnak Province [35].

Chemical analysis of Avgamasya asphaltites (%)								
Moisture	Ash	Volatile substance	Fixed carbon	C	H	N	S	O
0,40	47,28	34,30	18,20	42,1	3,57	0,5	5,55	47,88

Table 7.
Elemental analysis results of Avgamasya asphaltites [38].

presents the elemental analysis results of Avgamasya asphaltites [3, 30]. These properties enhance the potential use of asphaltite reserves in Şırnak for energy production.

3.2 Harbul fleet

The Harbul Field, located in the southern part of Mount Cudi, contains asphaltite veins distributed in lenticular levels within the Eocene-aged Gerçus formation. These veins are 1850 meters long and up to 55 meters wide. The total asphaltite reserve in this area is approximately 20,555,241 tons, with 12,704,241 tons considered visible reserves. According to evaluations by Turkey Coal Enterprises (TKİ), the reserve suitable for open-pit mining is estimated at 7,851,000 tons [3, 34]. Harbul asphaltites are dark in color, with approximately 10% pyrite content and 20% inorganic matter (**Table 8**). They contain quartz, calcite, dolomite, pyrite, gypsum, feldspar, mica, and clay minerals [3].

Chemical analysis of Harbul asphaltites	
Moisture (%)	1,77
Ash (%)	29,29
Volatile Substance (%)	55,42
Total Sulfur (%)	7,81
C (%)	56,49
H (%)	5
N (%)	0,75
O (%)	1,68
Lower calorific value (Kcal/kg)	5583
Upper calorific value (Kcal/kg)	5854

Table 8.
Properties and chemical analysis of Harbul asphalt [3].

3.3 Silip fleet

In order to evaluate the asphaltite potential in Şırnak Province, the data obtained by the drilling works carried out between 1982 and 1986 in the region between the Harbul fleet and the Silip fleet were examined. In this area, located 1 km east of the Harbul fleet, 22 soundings were carried out with a total depth of 5.112.70 m. In particular, the Silip fleet exhibits a different form of bedding than other fleets, and a total of 4,406,000 tons (visible reserve 3,071,000 tons and possible reserve 1,335,000 tons) were identified in this fleet [3, 34]. These results help us better understand the geographical distribution and potential of asphaltite sources in Şırnak Province.

3.4 Üçkardeşler fleet

Asphalt Distribution and Potential Assessment in Şırnak Province is based on detailed drilling work carried out on a fleet 4 km west of the Harbul fleet. This chapter includes 54 soundings carried out between 1980 and 1984, and these soundings were carried out in an area with a total depth of 10.584.10 m, shallow, upright, and oblique. In addition, with five inclined drilling made in 1985–1986, the total drilling depth of the fleet reached 11.527.50 m. According to the results of these comprehensive studies, a total of 19,775,159 tons (visible reserve 8.894.159 tons + possible reserve 10.881.000 tons) asphaltite reserve was determined in the fleet [3, 34].

3.5 İspindoruk fleet

Detailed research has been conducted on the presence of asphaltites in Şırnak province. In this context, according to the data obtained as a result of 11 splits and two oblique drilling studies carried out between 1972 and 1974, the distribution and potential of asphaltites in Şırnak Province were determined. As a result of these studies, 1.070.875 tons of asphaltite reserves were determined for a total depth of 100 meters. Visible in this reserve is classified as 70,875 tons, possible 500,000 tons, and possible 500,000 tons [3, 34].

3.6 Rutkekurat fleet

Deep drilling has not been carried out in the area on Mount Cudi in the northern region of Silopi. When a spread of approximately 800 meters and a depth of 50 meters were calculated with seven opening works in this region, the estimated asphaltite reserve was determined as 1.000.000 tons [39].

3.7 Uludere ortasu filonu

The fleet's research was initiated in 1982 with prospection studies. Successfully, in 1984, 12 splitting operations were carried out with a total of 12 upright and inclined drilling operations. Based on the results of these rigorous studies, it has been recorded that approximately 604,985 tons of asphaltite reserves have been identified in this particular region of Şırnak province [3, 34].

3.8 Milli fleet

Research on the fleet began in 1964 and continued periodically until 1974. During this time, 23 splitting operations and 10 inclined drilling were carried out on the fleet. The reserve of the fleet is determined as 6.500.000 tons. It is observed that the national fleet contains approximately 4% pyrite. Since the Rmax value is 0.38%, it can be concluded that the volatile content is low. It has also been found to contain around 13% clay and similar inorganic substances [35]. These analysis results are presented in detail in **Table 9**.

3.9 Karatepe fleet

Karatepe fleet was examined with 24 splits and four oblique soundings with a total depth of 307.30 meters within the scope of the research conducted between 1972

Chemical analysis of Mili asphaltites						
Moisture (%)	**Ash (%)**	**Volatile substance (%)**	**Total sulfur (%)**	**Fixed carbon (%)**	**Lower calorific value (Kcal/kg)**	**Upper calorific value (Kcal/kg)**
2,22	50,42	37,82	4018	9,54	2953	3084

Table 9.
Review of Mili asphaltites [35].

Chemical analysis of Karatepe asphaltites						
Moisture (%)	**Ash (%)**	**Volatile substance (%)**	**Total sulfur (%)**	**Fixed carbon (%)**	**Lower calorific value (Kcal/kg)**	**Upper calorific value (Kcal/kg)**
3,61	46,86	21,11	5,56	22,42	3760	3910

Table 10.
Analysis of Karatepe asphaltites [38].

Analysis of Seridahli asphaltites						
Moisture (%)	**Ash (%)**	**Volatile substance (%)**	**Fixed carbon (%)**	**Total sulfur (%)**	**Lower calorific value (Kcal/kg)**	**Upper calorific value (Kcal/kg)**
4,05	40,36	23,03	32,56	5,33	4406	4581

Table 11.
Analysis of Seridahli asphaltites [3].

and 1974. The estimated reserve amount of this fleet is determined as 4.522.096 tons. According to the results of the mineralogical analysis, this fleet contains calcite, dolomite, clay minerals, and pyrites, these results are presented in detail in **Table 10** [35].

3.10 Nivekara fleet

In the Nivekara fleet, possible and possible quantities of approximately 1,914,410 tons of asphaltite reserves up to 100 meters deep were detected in an area of 3300 meters, with a width ranging from 0.5 to 11.80 meters [3].

3.11 Segürek fleet

Segürek asphaltites have large pores and characteristically have a light gray color. The Rmax value was measured at 0.66% and its gaps are mostly filled with 14% pyrite. In addition, the proportion of clay and inorganic substances contained in these asphaltites is six [37].

3.12 Seridahli fleet

Seridahli fleet is located approximately 12 kilometers southwest of Şırnak province. The length of this vessel is 1750 meters and its width varies between 1.9 and 14 meters [16, 20]. Rmax value is recorded as 1.48% the pyrite content in this fleet is at 5% and the inorganic substance content is at 17% [34]. The estimated reserve amount of this special fleet is calculated as approximately 2,000,000 tons. Detailed analysis data of Seridahli asphaltites are presented in **Table 11.**

4. Use of asphaltites as an energy source in Şırnak

The lower calorific values (LCV) of asphaltites in the Şırnak nail basin range from 2876 to 5536 kcal/kg, with an average LCV of around 4500 kcal/kg (**Table 12**). While the moisture content is low, the ash and sulfur contents are high, making asphaltites unsuitable for domestic use. However, they are suitable for electricity generation, and the use of limestone in circulating fluidized bed (CFB) power plants can significantly reduce ash and sulfur emissions. The 405 MW capacity thermal power plant in the Silopi district serves as an important example of the use of asphaltites in energy production.

Fluidized-Bed Combustion (FBC) is a technology capable of burning low-quality fuels, rich in sulfur and ash, in an effective and environmentally friendly manner (**Figure 10**). It has been successfully applied in power plants since the 1980s. In FBC, solid particles are homogeneously moved by air from the bottom through a distributor

Energy properties of Şırnak asphaltites	
Fleet name	**Lower calorific value (kcal/kg)**
Harbul	5536
Silip	5485
Üçkardeşler	5474
Rutkekurat	3250
Avgamasya	4191
Mili	3400
Anılmış-Karatepe	3695
Seridahili	3174
Nivekara	3400
Segürük	4500
Anılmış-İspindoruk	3300
Uludere Ortasu	2876

Table 12.
Energy properties of Şırnak asphaltites.

Figure 10.
North face of Avgamasya No. 1 pit Şırnak asphaltite coal mine site [40].

plate, and coal burns perfectly in inert particles. Combustion takes place between 800 and 900°C, and during this process, SO_2 is captured with limestone without the need for an additional chimney. Additionally, small particles carried by the gas are separated by the cyclone at the outlet of the combustion unit and returned to the bed, extending the residence time and increasing combustion efficiency A symbolic representation of fluid bed technology is presented in **Figure 11** [42].

The advantages of FBC technology include high combustion efficiency, low NOx and SO_2 emissions, compatibility with low-quality fuels, and the use of ash in various industrial applications [39, 40]. It offers more efficient and environmentally friendly solutions for energy production.

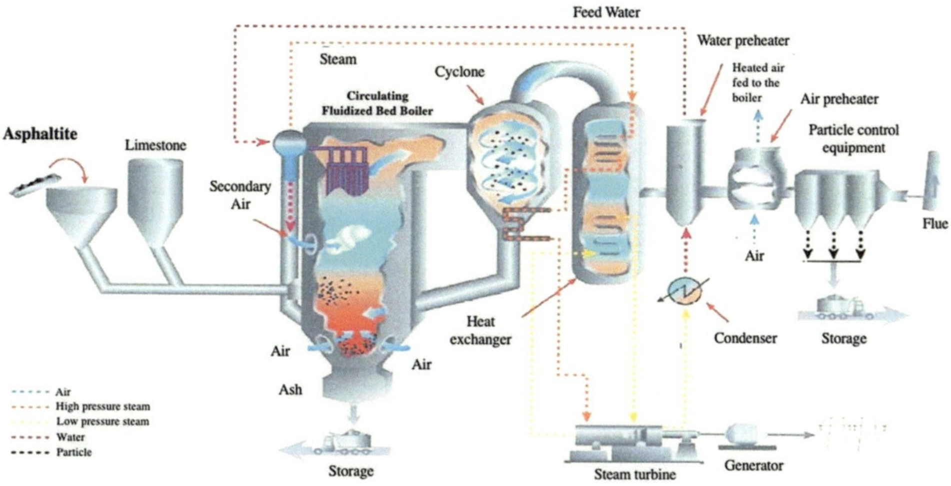

Figure 11.
Fluidized bed combustion (FBC) system flow diagram [41].

FBC system units consist of the following components [42]:

- Fluidized bed boiler and associated auxiliary facilities,
- Steam turbines and generators,
- Cooling water and related cooling systems,
- Water preparation facilities,
- Asphaltite and limestone preparation facilities,
- Ash disposal system,
- Electrostatic filters,
- Auxiliary equipment and administrative buildings.

5. Conclusions

This chapter investigated the distribution and potential of asphaltites in Şırnak Province. Asphaltite is a natural substance resembling asphalt, derived from oil and formed by factors such as hydrostatic pressure, gas pressure, gravity, and temperature. In the Şırnak region, asphaltites are found in liquid or semiliquid forms within faults and cracks. The key findings are as follows:

- Asphaltite reserves: Asphaltite reserves have been identified in Şırnak Province. This natural substance, formed from oil, is present as fault- and crack-filling asphalt in significant quantities, particularly in the Silopi Harbul fleet.

- Reserve amount: Studies conducted in Şırnak Province and Silopi district determined a total of approximately 82 million tons of asphaltite reserves, with the highest concentration (61.65%) located in the Silopi Harbul fleet.

- Obtaining minerals: Positive results were obtained from asphaltites in terms of rare minerals (V, Ni, Mo, U_3O_8, Ti), presenting a potential resource for the mining industry.

- By-product production: Ammonia and H_2S, by-products of synthetic oil from asphaltites, were also extracted, offering significant opportunities for the energy and chemical industries.

- Ash and sulfur content: Due to the high ash and sulfur content, asphaltites are unsuitable for domestic use. However, the use of limestone in circulating fluidized bed power plants can reduce ash and sulfur emissions.

- Energy potential: The asphaltites in Şırnak have high lower thermal values, with an average LCV of 4500 kcal/kg and above, making them suitable for energy production.

- Example application: The 405 MW thermal power plant in Silopi serves as a key model for utilizing asphaltite for energy production, highlighting the region's potential for asphaltite-based energy.

Conflict of interest

The authors declare no conflict of interest.

Thanks

We would like to express our sincere thanks to MTA (Mineral Research and Exploration General Directorate), TKİ (Turkish Coal Enterprises), and Şırnak University for their valuable contributions in providing the data for this chapter. MTA and TKİ strengthened the foundation of this research with their invaluable data and technical information, while Şırnak University made significant contributions through its academic support and resources. The contributions of these institutions have played a crucial role in the success of this chapter.

Appendices and nomenclature

AID	lower calorific value
CFB	circulating fluidized bed
CS_2	carbon disulfide
g/cm^3	gram per cubic centimeter
H	hydrogen
H_2S	hydrogen sulfide
kcal/kg	kilocalories per kilogram

KWh	kilowatt-hour
MJ/kg	megajoule per kilogram
Mo	molybdenum
MTA	mineral research and exploration
Mtoe	million tons of oil equivalent
MW	megawatt
Ni	nickel
NOx	nitrogen oxides
O	oxygen
Rmax	maximum reflectance value
S	sulfur
SO_2	sulfur dioxide
TKİ	Turkey coal enterprises
Ti	titanium
TWh	terawatt-hour
U_3O_8	uranium oxide

Author details

Edip Taşkesen[1*] and Hamza Alahmad[2]

1 Faculty of Engineering, Department of Energy Systems Engineering, Şırnak University, Şırnak, Türkiye

2 Energy Science and Technologies Program (MSc), Institute of Graduate Education, Şırnak University, Şırnak, Türkiye

*Address all correspondence to: edip.taskesen@şırnak.edu.tr

References

[1] Asif M, Muneer T. Energy supply, its demand and security issues for developed and emerging economies. Renewable and Sustainable Energy Reviews. 2007;**11**(7):1388-1413. DOI: 10.1016/j.rser.2005.12.004

[2] Demir M. The relationship between energy import and current account deficit: The case of Turkey with VAR analysis. Journal of Academic Researches and Studies. 2013;**5**(9):2-27. DOI: 10.20990/aacd.79126

[3] Demirci S, Sivrikaya O, Vapur H. Enerji kaynağı olarak asfaltit: oluşumu, içeriği, Türkiye rezervleri, temizlenmesi. Ömer Halisdemir Üniversitesi Mühendislik Bilimleri Dergisi. 2019;**8**(1):312-325. DOI: 10.28948/ngumuh.517131

[4] Davenport J, Wayth N. Statistical Review of World Energy: 2024. 73rd ed. London: Energy Institute; 2024

[5] Our World in Data Org. Primary Energy Consumption from Fossil Fuels, Nuclear and Renewables, World [Internet]. 2024. Available from: https://ourworldindata.org/ [Accessed: November 12, 2024]

[6] National Annual Estimate. The Results of Address Based Population Registration System. 2023. Available from: www.tuik.gov.tr

[7] World Bank Group. World Bank Group. World Development Indicators. Available from: https://databank.worldbank.org/source/world-development-indicators

[8] Our World in Data. Electricity Production by Source [Internet]. 2024. Available from: https://ourworldindata.org/

[9] World Energy Outlook 2023. Paris: IEA-International Energy Agency; 2023

[10] Aydın O. Çözüm süreci ve madencilik. Madencilik Türkiye Dergisi. 2013;**1**(31):82-95

[11] Kural O, Pişkin S. Asfaltitlerin değerlendirilmesi ve Konya ılgın kömürlerinde katkı maddesi olarak kullanılması. In: Türkiye 6. Kömür Kongresi. Zonguldak: TMMOB Maden Mühendisleri Odası; 1998. pp. 261-264

[12] Orhun F. Characteristic properties of the asphaltic substances in southeastern Turkey, their degrees of metamorphosis and their classification problems. Bulletin of the Mineral Research and Exploration. 1969;**72**(72):146-157

[13] İnsapedia. Asfaltit Nedir [Internet]. 2019. Available from: https://insapedia.com/ [Accessed: November 12, 2024]

[14] Kelso JL, Powell AR. Glance pitch from tell Beit Mirsim. Bulletin of the American Schools of Oriental Research. 1944;**95**:14-18

[15] Berkowitz N, editor. Fossil Hydrocarbons: Chemistry and Technology. California: Academic Press; 1997. 343 p

[16] Işıganer T. Mardin-Silopi-Harbol (Aksu) ve Üçkardeşler asfaltit filonlarına ait jeoloji raporu. Ankara: Maden Tetkik Arama Genel Müdürlüğü "MTA"; 1985. p. 92

[17] Nakoman E. Sedimanter uranyum yatakları ve Türkiye'nin bu yönden olanakları. Ankara: MTA Enstitüsü; 1977. pp. 1-28

[18] Parnell J. Timing of hydrocarbon-metal interactions during basin evolution. In: 25 Years Society for

Geology Applied to Mineral Deposits Anniversary Meeting. Netherlands. Aug [Online] 1991. pp. 573-576. [Accessed: Dec 31, 2024]. Available: http://pascal-francis.inist.fr/vibad/index.php?action=getRecordDetail&idt=6543757

[19] Parnell J. Hydrocarbons and other fluids: Paragenesis, interactions and exploration potential inferred from petrographic studies. Geological Society. 1994;**78**:275-291

[20] Lebkuchner RF, Orhun F, Wolf M. Asphaltic substances in southeastern Turkey. American Association of Petroleum Geologists "AAPG Bulletin". 1972;**56**(10):1939-1964. DOI: 10.1306/819A419E-16C5-11D7-8645000102C1865D

[21] Saltoglu T, Akyuz T, Alparslan E. Quantitative determination of molybdenum, nickel, vanadium and titanium in the asphaltites and asphaltite ashes by XRF-spectroscopy. Bulletin of the Mineral Research. 1978;**91**:89-93

[22] Kırşan İH. Doğu ve Güneydoğu Anadolu'nun Kalkınması İçin Önemi Olan Maden Yatakları. Ankara: Maden Mühendisleri Odası; 1993. pp. 1-3

[23] Kogel JE, Trivedi NC, Barker JM, Krukowski ST. Industrial Minerals & Rocks, Commodities, Markets and Uses. 7th ed. Colorado: Society for Mining, Metallurgy and Exploratio "SME"; 2006. 1507 p

[24] Jahanian HR, Shafabakhsh GH, Divandari H. Performance evaluation of hot mix asphalt (HMA) containing bitumen modified with gilsonite. Construction and Building Materials. 2017;**131**:156-164. DOI: 10.1016/j.conbuildmat.2016.11.069

[25] Boden T, Tripp BT. Gilsonite Veins of the Uinta Basin, Utah. Vol. 141. Utah Geological Survey; 2012

[26] Wikipedia. Asphaltite [Internet]. Aug [Online]. 2024. Available from: https://en.wikipedia.org/wiki/Asphaltite [Accessed: December 31, 2024]

[27] Ziyadanoğulları İAR. Leaching of asphaltite ash with H_2SO_4 and extraction of Mo, V, U in solution. Cankaya University Journal of Arts and Sciences. 2005;**1**(4):143-155

[28] Speight JG. Asphalt Materials Science and Technology. Waltham (USA): Butterworth-Heinemann is an imprint of Elsevier; 2015. p. 619

[29] Çakır B, Kılıç AM, Kahraman E. Open pit mining design of Silopi (Harbul-Uckardesler) asphaltite filon by micromine mining design software. Çukurova University Journal of the Faculty of Engineering and Architecture. 2013;**28**(2):99-112

[30] Ünalan G. Kömür jeolojisi. Ankara: Maden Tetkik ve Arama Genel Müdürlüğü; 2010

[31] TMMOB Enerji Raporu 2006. 1st ed. Ankara: Türk Mühendis ve Mimar Odaları Birliği; 2006

[32] Hiçyılmaz C, Emre AN. Improvements on combustion properties of asphaltite and correlation of activation energies with combustion results. Fuel Processing Technology. 2006;**87**(6):563-570. DOI: 10.1016/j.fuproc.2005.07.010

[33] Gönenç O. Asfaltitler ve Türkiye'deki asfaltit yatakları. Ankara: MTA Enstitüsü Dergisi; 1990

[34] Şengüler İ. Asfaltit ve bitümlü şeylin Türkiye'deki potansiyeli ve enerji değeri. In: EMO Türkiye VI Enerji Sempozyumu–Küresel Enerji Politikaları ve Türkiye Gerçeği; 22-24 Ekim 2007. Türkiye. Ankara: TMMOB EMO; 2007. pp. 86-95

[35] Şengüler İ. Ülkemiz enerji bütünlemesinde Marmara ve Trakya bölgesi kömürlerinin yeri. In: Marmara Enerji Forumu; 7-8 Eylül 2007. Türkiye, Ankara: TMMOB EMO; 2007. pp. 1-7

[36] Karayigit AI, Querol X. Mineralogy and elemental contents of the Sirnak asphaltite, Southeast Turkey. Energy Sources. 2002;**24**(8):703-713. DOI: 10.1080/00908310290086662

[37] Kavak O, Connan J, Erik N, Yalcin MN. Organic geochemical characteristics of Sirnak asphaltites in Southeast Anatolia, Turkey. Oil Shale. 2010;**27**(1):58-83. DOI: 10.3176/oil.2010.1.07

[38] Tosun Y. Rock sliding risk of Şırnak asphaltite coal quarry-study on cemented cracks of shale, marly claystone for upper slope's stability by sawing disc tests. EJONS International Journal. 2024;**8**(2):260-269. DOI: 10.5281/zenodo.12515720

[39] Hamamci C, Kahraman F, Düz MZ. Desulfurization of southeastern Anatolian asphaltites by the Meyers method. Fuel Processing Technology. 1997;**50**(2-3):171-177. DOI: 10.1016/S0378-3820(96)01077-6

[40] Chilingarian GV, Yen TF. Introduction: Organic matter and origin of oil and Tar Sands. Developments in Petroleum Science. 1978;7:1-15. DOI: 10.1016/S0376-7361(08)70060-8

[41] İli Ş, İlçesi Görümlü-Çalişkan S, Çed B, Nihai R, Raporu Ç. Şirnak-silopi termik santrali, santrala yakit sağlayan asfaltit sahasi ve kireçtaşi sahalari kapasite artişi projesi çed raporu [Online]. Available from: https://www.stgm.org.tr/sites/default/files/2020-09/sirnak-silopi-termik-santrali-santrala-yakit-saglayan-asfaltit-sahasi-ve-kirectasi-sahalari-kapasite-artisi-projesi-ced-raporu.pdf [Accessed: December 31, 2024]

[42] Selçuk N. Akışkan yatakta yakma teknolojileri. In: Türkiye'de Kömür Politikaları ve Temiz Kömür Teknolojileri Sempozyumu; 21-22 Ekim 1999. Türkiye. Akara: TMMOB; 1999